10 0684594 1
TINGHAM
AWN
FROM THE LIBRARY

FUNGAL BIOLOGY IN THE ORIGIN AND EMERGENCE OF LIFE

The rhythm of life on Earth includes several strong themes contributed by Kingdom Fungi. So why are fungi ignored when theorists ponder the origin of life?

Casting aside common theories that life originated in an oceanic primeval soup, in a deep, hot place, or even a warm little pond, this is a mycological perspective on the emergence of life on Earth. The author traces the crucial role played by the first biofilms – products of aerosols, storms, volcanic plumes and rainout from a turbulent atmosphere – which formed in volcanic caves 4 billion years ago. Moore describes how these biofilms contributed to the formation of the first prokaryotic cells, and later, unicellular stem eukaryotes, highlighting the role of the fungal grade of organisation in the evolution of higher organisms. Based on the latest research, this is a unique account of the origin of life and its evolutionary diversity to the present day.

DAVID MOORE is an Honorary Reader in the Faculty of Life Sciences at the University of Manchester. Having recently retired after 43 years researching and teaching genetics and mycology, his ongoing research activities include computer programs simulating fungal growth and differentiation, and genomic data mining. In recent years he has created the educational websites www.fungi4schools.org (sponsored by the British Mycological Society) and www.davidmoore.org.uk. He is co-author of *21st Century Guidebook to Fungi* (Cambridge, 2011).

D1614945

FUNGAL BIOLOGY IN THE ORIGIN AND EMERGENCE OF LIFE

David Moore

Faculty of Life Sciences, University of Manchester, UK

UNIVERSITY OF NOTTINGHAM
GEORGE GREEN LIBRARY OF
SCIENCE AND ENGINEERING
WITHDRAWN
FROM THE LIBRARY

CAMBRIDGE
UNIVERSITY PRESS

CAMBRIDGE UNIVERSITY PRESS
Cambridge, New York, Melbourne, Madrid, Cape Town,
Singapore, São Paulo, Delhi, Mexico City

Cambridge University Press
The Edinburgh Building, Cambridge CB2 8RU, UK

Published in the United States of America by Cambridge University Press, New York

www.cambridge.org
Information on this title: www.cambridge.org/9781107652774

© Cambridge University Press 2013

This publication is in copyright. Subject to statutory exception
and to the provisions of relevant collective licensing agreements,
no reproduction of any part may take place without
the written permission of Cambridge University Press.

First published 2013

Printed and bound in the United Kingdom by the MPG Books Group

A catalogue record for this publication is available from the British Library

Library of Congress Cataloging-in-Publication Data

Moore, D. (David), 1942–
Fungal biology in the origin and emergence of life / David Moore, Faculty of
Life Sciences, University of Manchester.
pages cm
Includes bibliographical references.
ISBN 978-1-107-65277-4 (Paperback)
1. Fungi–Evolution. 2. Life–Origin. I. Title.
QK604.2.E85M66 2013
571.5′92–dc23
2012029375

ISBN 978-1-107-65277-4 Paperback

Cambridge University Press has no responsibility for the persistence or
accuracy of URLs for external or third-party internet websites referred to
in this publication, and does not guarantee that any content on such
websites is, or will remain, accurate or appropriate.

/006845941

CONTENTS

ONE

LEARNING FROM LIFE ON EARTH IN THE PRESENT DAY

The only known habitat for life is the planet we call Earth. The only place we know the life experiment has been carried out is this planet Earth. We do not know how life originated here, however. Nevertheless, we understand a great deal about the physical and chemical conditions, the environments, and some of the spontaneous ('self-organising') mechanisms that the physics and chemistry of this Universe make possible, so there is no difficulty in formulating reasonable models for the emergence and onward evolution of living things. One of the best existing books about this topic starts like this:

> The main assumption held by most scientists about the origin of life on Earth is that life originated from inanimate matter through a spontaneous and gradual increase of molecular complexity. This view was given a well-known formulation by Alexander Oparin [Oparin, 1957a], a brilliant Russian chemist who was influenced both by Darwinian theories and by dialectical materialism. A similar view coming from a quite different context was put forward by J. B. Haldane [Haldane, 1929]. By definition, this transition to life via prebiotic molecular evolution excludes panspermia (the idea that life on Earth comes from space) and divine intervention. (Luisi, 2006, chapter 1, p. 1)

Although I intend to discuss here the notion of panspermia (see Chapter 5, below), I will not discuss the other mechanism that Pier Luigi Luisi says is excluded by prebiotic molecular evolution in the quotation above, namely divine intervention. I choose not to include this because personally I see no need to invoke divinely magical or mythological processes in the *scientific* story I wish to tell: basic physics and chemistry are enough. If you want to round out your reading on the topic then I suggest you start with chapter 1 in Luisi (2006) and chapter 6 of Lurquin (2003). For the real hard-core discussion you don't need to go much further than Dawkins (1986, 2006) and Scott (2009). If you put the words 'dawkins' and 'god' together in Amazon's search window, the software will display publications on both sides of the argument, and from the number of items offered you may get an inkling of why I decided not to venture into this part of the arena!

Having indicated what I will not discuss, let me indicate what I will describe here. This book examines the most likely mechanisms by which life arose and progressed on Earth as we understand it from the most recent research. In many ways this book is a tourist guide in which I will escort you, the reader/tourist, through more than 4 billion years of the historical journey experienced by this planet. Because the only known habitat for life is planet Earth, we will start by examining how the habitat that has cradled the life we know about was formed. Among others, we will visit the following intellectual places:

- The formation of the Earth, in context of the origin of the Solar System, and how the basic chemistry, physics and geology of the early Earth influence the search for the origins of life.
- From this argument I will advance the view that the origin of life may be a logical inevitability of chemical evolution in a wide range of environmental conditions and I will account for the origin of biological building blocks on a prebiological world from a wide range of sources.
- Although it is most frequently argued that the primitive Earth's ocean was the one and only location in which prebiological chemical evolution could have occurred, I will develop the notion that for very extensive periods of time, billions upon billions of aerosol droplets existed all over the protoplanet Earth within which billions upon billions of different chemical reactions *might* have taken place, forming molecules and reaction trains that *might* have contributed to the origin of life.
- This leads naturally to speculations about evolution of almost-living protocells of all sorts: those without a genetic apparatus, those comprised of autocatalytic chemical cycles, the RNA-world and protein-world precursors of living things, and those exploiting advantages of existing within lipid envelopes –with all of these existing at the same time but in different aerosol droplets.
- Bring those aerosols down to earth in rain or spindrift and they will form slime upon the sterile solidified volcanic lava. Where that slime is protected from hostile solar radiation by tephra or in lava bubble caves something like a prebiological biofilm will form, within which all

pathways were effective (if they *could* happen, they *did* happen) and where they coexisted and were coextensive they could work together, creating ever more integrated and interconnected chemical systems.

- Life emerged from non-enzymatic systems of autocatalytic chemical cycles which later acquired the capacity of enzymatically controlled metabolism.
- The characteristics expressed by the earliest living, or even pre-living, things (their 'lifestyle') were those that today would be categorised as being ***heterotrophic***. The first chemical machines used readily available nutrients. As these were exhausted, selective advantage was gained by those chemical machines able to release chemical catalysts into the environment to degrade the tars and other polymers accumulated by hundreds of millions of years of abiotic chemical reactions.
- Much of current biology is dependent on symbiosis, so 'symbiotic associations' *must* have also been important in prebiology.
- This will lead logically to the idea that different types of protocells, each good at one particular process, must have coexisted. Some might have found selective value in collaborating, or in symbiotically absorbing and using other protocells, or in exuding factors (primitive enzymes and/or toxins) that targeted other protocells to open them up and make use of their constituents.
- The first living things arose on Earth as soon as conditions allowed and within a relatively short time the Earth became a planet dominated by primitive bacteria, with the majority still in biofilms.
- Further levels of collaboration within those biofilms allowed the higher organisms (eukaryotes) to emerge, initially as a one-celled organism that possessed a mixture of characteristics from which evolved the animal, fungal and plant lineages that we know today.

There are a great many publications in the literature that deal with these topics, particularly among those published in the last ten to twenty years. So, why add yet another one? Well, I would claim that the unique feature of the argument I wish to present here is that it is based on appreciation of the central role of the fungal grade of organisation in the evolution of higher organisms. This has never been done as far as I am aware, which is a pretty remarkable omission, but I believe the main reason for this is that most writers on the topic of the origin and early evolution of life have been, and still are, essentially ignorant of the distinctive features of fungal cell biology that set them apart from plants and animals, with little more knowledge of fungi than that revealed to them in primary school.

And fungi are very different from both plants and animals. Despite this, for most of the early period of the science of biology, the fungi were thought to be plants, peculiar and non-photosynthetic perhaps, but plants nevertheless. Generally speaking 'fungi' in this context means mushrooms and toadstools, and the fact that they appear on the ground among green plants in nature accounts for their association with plants. Indeed, fungi were initially firmly classified as plants and were even given an evolutionary history that derived them from algae by loss of photosynthesis. In the early part of the twentieth century it became increasingly clear that there are a great many fungi in the world. Beyond the most obvious mushrooms and toadstools, which are themselves incredibly diverse and numerous, there are enormous numbers of filamentous moulds: microscopic and soil fungi that have colonies, called mycelia, where the main body of the fungus is made up of thin, cottony, thread-like filaments that are called hyphae. There are also a great many yeasts: single-celled organisms that are specialised for growth in liquids, particularly small volumes of liquids like the sugary fluid in flower nectaries, or raindrops and dew drops on fruits and flowers, as well as the fluid circulation streams of both animals and plants in parasitic or pathogenic species. Yeasts are perfectly respectable fungi, though their specialism to a particular habitat has made them reduced in their form and structure, so their relationship to the overwhelming majority of other fungi, which are characteristically filamentous (or 'hyphal'), is in many respects similar to the relationship of flightless birds to the majority of other birds.

Most books about the origin of life have been written by physicists, cosmologists, astronomers and molecular biologists, all of whom, for some reason, think that biology is a lesser science than their own. For example: 'physics and chemistry, possibly more than biology, held clues to [the origin of life and the cosmos itself]', which is a quotation from the first paragraph of the preface to Lurquin (2003), who is a plant chemist/molecular biologist. Other authors apologise for their shortcomings but proceed undaunted; for example, the physicist Dyson (1999) quotes this from Schrödinger (1944) (yes, the one with the cat):

> ... some of us should venture to embark on a synthesis of facts and theories; albeit with second-hand and incomplete knowledge of some of them, and at the risk of making fools of themselves. So much for my apology.

And Freeman Dyson adds:

> This apology for a physicist venturing into biology will serve for me as well as for Schrödinger, although in my case the risk of the physicist making a fool of himself may be somewhat greater. (Dyson, 1999, chapter 1, p. 2)

A properly modest standpoint, except that, save for photosynthesis, Dyson's view of biology is narrowed to include only animals, which leads him to worry about aspects like encephalopathic prions and embryonic development of animals; neither of which have much relevance to the origins of life.

Similarly, while there is a great deal to admire in the chemoton theory of the chemical engineer Tibor Gánti (Gánti, 2003) he is extremely animal-centred, as is Conway Morris (2003); and Lane (2010) reveals his bias with the phrase 'so many creatures, even plants, indulge in sex' (chapter 1, p. 3).

Animal centricity is one of the two common failings of many of the 'origin of life' publications: plants often get a mention, but a strong impression remains that they are mentioned only because their photosynthesis provides the essential oxygen for the important organisms, animals, to get on with the business of being real living creatures. James R. Griesemer (in Gánti, 2003) refers to the 'vertebrate bias' with these words:

> Hull (1988) pointed out that there is a substantial 'vertebrate bias' in the types of features that biologists, who are all vertebrates, take to be 'typical' of organisms, even though vertebrates are very 'atypical' organisms (see also Buss 1983, 1987). Modern genes may be just as atypical of replicators in general as vertebrates are of organisms in general; it is an empirical question with implications ... (Gánti, 2003, chapter 5, p. 172)

It's worth noting that even Charles Darwin was close to suffering this animal bias in his work on the insectivorous higher plants known as sundews (Latin generic name *Drosera*). His wife, Emma Darwin, in a letter to Lady Lyell, August 1860, said:

> At present he [Charles] is treating *Drosera* just like a living creature, and I suppose he hopes to end in proving it to be an animal. (Darwin & Litchfield, 1915, p. 177)

Just like a living creature? Is a plant not a living being? Many of the papers and books that have been published recently, as well as the majority of biology and natural history programmes broadcast on radio, TV and Internet reinforce this animal bias.

The second common failing is that fungi, if mentioned at all, are dealt with in a perfunctory and sometimes indifferent manner. In Tom Fenchal's book, *The Origin & Early Evolution of Life*, the Introduction starts promisingly enough:

> This book is about the development of life from its origin and until multicellular plants, fungi, and animals arose ... (Fenchel, 2002)

Unfortunately, fungi are acknowledged to be one of the crown groups of eukaryotes in this way:

> Plants, animals and fungi are relatively closely related, but they derived independently from different protists. The fungi are more closely related to animals than to plants (which should be noted especially by botanists and vegetarians). (Fenchel, 2002, chapter 12, p. 122)

The recognition is satisfying, and is more than most books manage, but I find the closing phrase unnecessarily dismissive.

I respect these various sources for the contributions they have made to discussion of the origin of life, and (as you must have realised by now) I will be quoting from them where appropriate, with all due reference and acknowledgement, even though I consider their lack of appreciation of fungi a severe disappointment. Their understanding of the physical sciences may be enviable, but their appreciation of biology and living systems is frequently naive and sometimes banal. It seems that the biologists too, even the eminent ones, are no better informed and I trace the cause to severe flaws in our educational system. As I have commented before, with various colleagues (Moore & Põder, 2006; Moore *et al.*, 2006; Moore, Robson & Trinci, 2011, preface, pp. ix–x) this partial view of biology afflicts the school curriculum and results in most current biology teaching, from school-level to university, concentrating on animals, with a trickle of information about plants and bacteria. This results in the majority of school and college students (and, since they've been through the same system, most current university academics) being ignorant of fungal biology and of their own dependence on fungi in everyday life. Allow me to explain.

I think it's likely that the lack of a proper appreciation of the fungal lifestyle is a severe limitation to understanding the origin and early evolution of life on Earth because, as I will argue below, the first eukaryotes had the nutritional and cell-biological attributes of fungi as well as features that later emerged in plants and animals. From this primitive, almost protofungal, stem the first to diverge were the plants. This left what are known as opisthokonts, ancestors of both animals and fungi, which are organisms with a single posterior flagellum. Finally the animals diverged from these, leaving their sister ancestral opisthokonts to evolve into fungi. That is, animals are the last of the crown group of eukaryotes to diverge and are therefore least informative about the point of origin of eukaryotes; and although they diverged early, plants are narrowly channelled into a specific way of life and are therefore less informative than fungi about the point of origin. Known fungal evolution 'is not marked by change and extinctions but by conservatism and continuity' (Pirozynski, 1976); an observation comparing very ancient fossil fungal spores to modern fungi, which implies that the fungal lifestyle will be informative about the point of origin. I should explain here that I am using the word 'lifestyle' as a synonym for 'body plan' as used by Cavalier-Smith (2006, 2010a):

> It cannot be coincidental that the largest expansion of protist diversity in Earth history immediately followed these global glaciations. The pump was primed by the earlier origin of eukaryotes. Glacial melting did not initiate cellular innovation; it just released the pent-up potential for

> innovation and rapid radiation that major new body plans themselves create. (Cavalier-Smith, 2010a, p. 127)

It is clear that Tom Cavalier-Smith includes physiological features under the same phrase:

> ... adaptive zones for the new phyla were created for the first time by their novel body plans, e.g. water splitting (oxygenic) photosynthesis by cyanobacteria and their immediate ancestors, phagotrophy by eukaryotes. (Cavalier-Smith, 2006, figure 8 legend)

Indeed, the fungal nutritional lifestyle/body plan has been at the essential centre of the evolution of life on this planet from its very start. To remind ourselves about this it is worth extracting some quotations from Robert Harding Whittaker's original 'New concepts of kingdoms of organisms' paper (Whittaker, 1969). With regard to nutritional modes, Whittaker wrote:

> There are, however, not two principal modes of nutrition [photosynthetic and ingestive] but three – the photosynthetic, absorptive, and ingestive. The three modes largely correspond to three major functional groupings in natural communities, the producers (plants), reducers (saprobes, that is, bacteria and fungi), and consumers (animals) ... The importance of the reducers in the cycling of materials in ecosystems appears to exceed that of the consumers. In evolution ingestive nutrition was a development secondary to the absorptive nutrition of most monerans and many eucaryotic unicells. Both protozoans with food vacuoles and metazoans with digestive tracts have probably evolved from absorptive flagellates, and in this evolution internalized the process of food absorption and added to it the process of ingestion. One may consider that the eucaryotic plants also have internalized the absorption of food through a membrane, that surrounding the chloroplast as symbiont and organelle. The three modes of nutrition imply different logics on which the evolution of structure in higher organisms was based ... (Whittaker, 1969, p. 152)

The relevance of this to discussion of the origin of life is that even in discussions of prebiotic chemistry a common notion is that some sort of compartment would absorb material from its outside to convert to more of the substance of its inside (an 'absorptive protonutrition').

> The taking in of organic substances dissolved in the surrounding aqueous medium and their transformation into parts of its own body is, obviously, the absolutely indispensable form of metabolism in a living body which arises by the incorporation of polymeric organic compounds into multimolecular systems. (Oparin, 1957a, chapter IX, p. 400)

I will show below that there is good reason to believe that the prebiotic primeval Earth was well supplied with a wide range of organic compounds varying from simple sugars and amino acids to peptides, carbohydrate polymers and polyaromatic hydrocarbons (PAHs) made up of several to many rings of carbon atoms – all these being the results of abiotic synthesis in locations as different as the surface of the Earth itself or other objects in the Solar System, through to interstellar space. Most arguments see the need for the prebiotic compartment (Oparin's coacervate; see Chapter 9) to first take up simple compounds from its medium and then, as these are exhausted, to release catalytic materials (reactive ions or metabolites/peptides/ribozymes) into the medium to break down larger molecules into simple compounds that the compartment can continue to absorb.

This is a step on the way to life; a prebiotic stage which depends on an evolutionary logic (using the phrase coined by Whittaker, 1969) that is the principal nutritive mode (saprotrophism) among modern-day bacteria and fungi; and it is the shared evolutionary logic which is the important point. The fungal grade of cellular organisation is a eukaryotic grade of organisation, but the strands of evolutionary logic that gave rise to it apply just as much to the primeval (prebiotic) period as to the later periods of cellular evolution of prokaryotes and eukaryotes alike. It is perhaps worth mentioning that despite recent antipathy to use of the word 'prokaryote' the prokaryote–eukaryote dichotomy 'reflects a profound evolutionary truth' (see discussion in Cavalier-Smith, 2010a, p. 113) and is one that I will continue to use.

Fungi, which today form a gigantic and diverse group of organisms, were all classified as plants right up to the middle of the twentieth century. This is when the notion began to crystallise that fungi might form a group of higher organisms different from both plants and animals. That is, that in their own right fungi might form a distinct kingdom of eukaryotes (organisms whose cells contain complex membrane-surrounded structures, in contrast to the mainly bacterial prokaryotes that generally do not have membranous structures within their cells). A classic publication on this topic is the paper, from which I have already quoted, entitled 'New concepts of kingdoms of organisms' (Whittaker, 1969) that contributed, with many other publications, to the present situation where fungi are not classed as animals or plants, but have a kingdom of their own to which they belong. This 'Kingdom Fungi' is amazingly diverse and includes organisms that range from being just a single cell, like the yeasts, to others that cover hundreds of acres of land. Indeed, some fungi are among the largest organisms on the planet (Arnaud-Haond *et al.*, 2012). One individual mycelium of the tree pathogen *Armillaria ostoyae* has been found in the mixed-conifer forest of the Blue Mountains of northeast Oregon in the USA that covers 965 ha (2384 acres; equivalent to 1355 Premiership football fields, or seven times the size of London's Hyde Park and three times the size of New York's Central Park). This mycelium is estimated to be 1900 to 8650 years old (Ferguson *et al.*, 2003).

To date, 100 000 species of fungi have been discovered but it is thought that there are over one and a half million species still to be found. Fungi can live in many habitats from polar regions to temperate and tropical forests, and in both fresh and salt water. However, most fungi live in soil. The fungi that most people are familiar with are those that form mushrooms and toadstools (there's no technical difference between a mushroom and a toadstool, essentially the words are synonyms). But mushrooms are not organisms in their own right, they are the fruiting bodies of a much larger and more extensive mycelium growing beneath the ground within the soil (mushrooms are analogous to the apples on an apple tree). The Kingdom Fungi is extremely diverse; in the British Isles there are 15 000 known species of fungi, 4500 of which are mushrooms (200 edible and only about 50 being poisonous to some degree), yet the British Isles can claim only 48 known species of mammal, 210 species of birds, and 1500 species of higher plant. The number of fungal species in Britain rivals the known species of insects (22 500 British species), but that comparison might misleadingly underestimate the diversity of the fungi as virtually every individual insect carries a load of parasitic fungi, many of which are species-specific, so in Britain (and the rest of the world) there may be many as yet unnamed entomogenous (which means 'growing on or in the bodies of insects') fungi.

Fungi are not able to produce their own food directly as plants do; rather, fungi recycle dead organic matter by discharging a full range of digestive enzymes into their surroundings. These enzymes degrade organic material outside the mycelium and then the hyphae ingest the soluble nutrients produced by that external digestion. Organisms like this are said to be saprotrophs, which means 'decomposer' because they live by digesting debris left by other organisms. A saprotroph is a specific kind of heterotroph, or consumer, which gets its nutrients ready-made from its environment. This distinction will become important when we have to start thinking about how the first organisms made a living a few billion years ago. Those that were literally 'the first' must have got their nutrients from their surroundings, but by definition, those nutrients could not have been left as debris by other organisms (because there hadn't been any); so the first organisms were heterotrophs, just consuming what was available (and as we will see, quite a wide range of organic compounds was available). But as soon as some of those organisms died, the second generation of organisms that came along would have been able to evolve ways of using the dead debris of the first; that is, they could have been saprotrophs, recycling debris like modern-day fungi. They would not have been fungi in those distant days, but they would have shared the characteristic fungal lifestyle.

In most soil environments of the present day, the bulk of the debris is made up of plant litter, such as leaves, plant stems and woody branches. But fungi are able to produce a complete range of digestive enzymes and are equally able to digest the protein, cartilage and bone of animal cadavers, and the chitinous exoskeletons of insects. They can also extract nutrients, like metal ions, phosphorus and sulphur, from inorganic minerals. But it is in the digestion of woody plant cell walls that the fungi demonstrate their unique abilities to recycle. Only fungi can degrade woody lignin, and without fungal wood decay the world would fill up with dead timber – as it did in Carboniferous days, 360 million years ago (mya), when the thickest coal seams were laid down because at that time the fungi had not yet evolved the ability to recycle dead timber so the wood just accumulated until it fossilised into coal.

The cell walls of plants are very strong. The components that provide the strength are cellulose (the sugar-based molecule from which the first, or primary, plant cell wall is made) and lignin (the phenol-based molecule that is characteristic of the plant cell's secondary cell wall). Fungi are very important for the decay of wood because they are the only organisms capable of breaking down both cellulose and lignin. Cellulose is a polymer of glucose that forms fibres that are incredibly strong. Brown rot fungi break down cellulose. Brown rot fungi are so called because the lignin remains intact and the wood keeps its brown colour. The enzymes released by brown rot fungi break the cellulose chains into single molecules of the sugar glucose that can be absorbed and used by the fungus. Lignin is the other strong polymer. It is the second most abundant natural polymer on Earth after cellulose. Fungi that break down lignin are called white rot fungi; this is because as it is digested and the content of lignin is decreased, the wood becomes lighter in colour. White rot fungi degrade lignin by producing oxidising enzymes that are released from their hyphae; in a very real sense they 'burn' the wood in an enzyme-controlled way. Lignin contains phenols and the white rot fungi are the only organisms that can deal with them (the fungi open up the carbon rings to form open chain molecules that can be metabolised). These two types of fungi have important roles in the recycling of nutrients. Without them, old plant material would not decay and the soil nutrients would be locked into an accumulating mass of undegradable lignin-based biomass. So, today, it is entirely thanks to the fungi that all the leaves that fall from the trees, the branches, and even the whole dead trees that fall to the ground in storms are broken down and recycled into nutrient-rich humus that can be used by plants for their growth.

This lifestyle, based on their ability to exude digestive enzymes that are able to recycle most of the organic (and some of the inorganic) matter in their environment, is a major contribution that fungi make to life on this planet, but it is not their only contribution. Fungi form a crucial part of the food web in most natural habitats; snails and slugs, mice, voles, squirrels and deer regularly eat lichens, mushrooms, toadstools and truffles as a major part of their diets, while millipedes, insects, insect larvae and beetle grubs eat the hyphae of mycelia in the soil. Throughout their existence, fungi have formed 'alliances' with other organisms in ways that enabled the joint or co-operative organisms to thrive in conditions that neither single organism could easily tolerate. In biology such an alliance used to be called a symbiosis, but it's more usual today to call it a mutualism because this term immediately indicates the fact that both organisms enjoy a mutual benefit.

Probably the most ancient mutualism is that between filamentous fungi and photosynthetic unicellular algae to form the lichen symbiosis. The first of these involved cyanobacteria, those photosynthetic bacteria which were the first organisms to release oxygen into the Earth's atmosphere. The most ancient lichen-like fossils so far reported were found in South China. These fossils involve filamentous hyphae in close association with cyanobacteria and are between 551 and 635 million years old. They show that fungi developed partnerships with photosynthetic organisms well before the evolution of vascular plants (Yuan, Xiao & Taylor, 2005). Fossil lichens have also been described from the Rhynie Chert which is 400 million years old (Taylor, Hass & Kerp, 1997). As the more advanced (eukaryotic) algae evolved, the fungi (mostly relatives of present-day ascomycete moulds) developed the partnerships that are now lichens, which are usually associations between a fungus and a green alga, though cyanobacterial lichens do persist.

The basis of the association is that the alga (known as the photobiont) uses sunlight to produce carbohydrate photosynthetically that is shared by both partners in the lichen. Some lichens contain both green algae and cyanobacteria as photobionts, in which case the cyanobacterium may focus on fixing atmospheric nitrogen for the joint metabolism of this tripartite association. But the algal partner(s) constitute(s) only 5% to 10% of the total biomass of the lichen; most of the thallus is fungus, which must be considered the dominant partner that 'construct[s] a plant-like body within which photosynthetic algal symbionts are cultivated' (Sanders, 2001); this quotation clearly suggests that the fungus is 'farming' its photobiont. In the present day about 20% of all known fungi are involved in lichens.

The lichen tissue formed by the fungus (called the thallus) is very different from either a fungal mycelium or an alga growing separately. Fungal hyphae surround the algal cells, often incorporating them into multicellular fungal tissues unique to lichen associations. There are about 20 000 species of these unique composite organisms today; they vary in size, shape and colour. Some are flat and firmly attached to the surfaces on which they grow, like those yellow/orange/brown discs that are often seen scattered over walls and roofs. But others are scaly, leafy or bushy, or hang in strands from their supports; some lichen thalli are very similar to simple plants in appearance and growth.

The lichen association is an intimate symbiosis which extends the ecological range of its partners; today's lichens dominate more of the land surface of the Earth than do tropical rainforests, and are able to live in places that are inaccessible to other organisms. A characteristic feature is that lichens can tolerate severe desiccation. Dry lichens can survive extremes of temperature, irradiation and other harsh environments. Lichens are pioneers, invading places where there's nothing more than rock and mist to live on. The fungus protects the alga and supports it physically and physiologically by taking in water and using its externalised enzymes to extract nutrients from whatever soil there might be, and even from the rocks themselves, leading to the mobilisation and absorption of minerals like magnesium, manganese, iron, aluminium and silicon. The algal and/or cyanobacterial cells in the partnership are able to photosynthesise, photolysing water and reducing atmospheric carbon dioxide to provide carbohydrates for use by both partners; cyanobacteria can also fix atmospheric nitrogen for the partnership. Just how mutual the benefits might be are in doubt, however. Microscopic examination shows that fungal cells might even penetrate the algal cells in a way similar to pathogenic fungi, although for the most part the readiness with which nutrients leak out of algal cells makes penetration by fungal hyphae unnecessary. Cells of the photobiont are routinely destroyed in the course of nutrient exchange, though, stability of the association depending on photobiont cells reproducing more rapidly than they are destroyed. Such observations suggest that the fungus in the lichen is effectively parasitising the alga, and using the products of algal photosynthesis for fungal nutrition.

For reproduction several lichens produce flakes made up of small groups of algal cells surrounded by fungal filaments (called soredia). Other reproductive structures are fragile upright elongated outgrowths from the thallus (called isidia) that break off for dispersal. Many other lichens break up into fragments when they dry, being dispersed by the wind, to resume growth when moisture returns. In any of these cases the fragments include both fungus and alga so the composite organism is reproduced intact. In addition, fungal sexual structures are often formed in abundance, though the fungal spores they produce must find a compatible algal partner before a functional lichen can be formed.

Plants gain their nutrients by absorbing minerals and water from the soil using their roots. But a great many plants get quite a lot of help from certain species of fungi. The relationship appears to have started because the plant roots alone are not able to supply the plant with all the nutrients it needs. The fungi associated with plant roots are called mycorrhizas, the hyphae of which are intimately associated with plant root cells and also extend into the surrounding soil to increase nutrient availability to the plant. The numerous hyphae of the fungi greatly increase the surface area available for absorbing minerals. The hyphae can also go looking for food, by growing into areas of fresh nutrients. The relationship between the plant and fungus is mutualistic.

Both sides gain something from having the other present; the plant pays for the privilege of using this fungus to bring it nutrients by sharing up to 25% of the products of its own photosynthesis with the fungus. The fungus benefits by taking readily available sugars from the plant. Despite this 'tax' on its activities, the plant grows much better than it would without the mycorrhiza. Some mycorrhizal fungi form a mat of fungal tissue around the root; the fungal cells grow between the cells of the plant root, but never actually cross the plant cell walls. These are called 'ectomycorrhizas'. In another mycorrhizal partnership (called an endomycorrhiza) the fungal cells enter the plants cells. Inside the plant cells they make structures that exchange nutrients with the plant cytoplasm. By greatly increasing the absorbing surface of a host plant's root system, mycorrhizas improve the plant's ability to tolerate drought and other extremes, such as extremes of temperature and acidity. As many as 95% of all plants have mycorrhizal associations, showing just how important these types of fungi are for the growth of so many plants, including all the crop plants we need to feed the human population, and all the trees in all the forests. More than 6000 fungi are capable of forming mycorrhizas. Mycorrhizal associations originated over 600 million years ago and developed very early in colonisation of the terrestrial environment. So they helped to create the world we have today.

Some important animals are also in mutualistic associations with fungi. Many animals including cows, sheep, goats, deer, and even giraffes, are known as ruminants. This is because they have a specialised four-chambered stomach needed for the digestion of their exclusively vegetarian diet. The first chamber the food enters is called the rumen, hence the name ruminant. The ruminant we're most familiar with is the cow, and we all know that cows spend most of their time eating grass and hay. Plant cell walls contain cellulose, which is an excellent source of fibre in the diet of most animals. Fibre is important as it provides roughage which keeps the excretion of waste products regular. However, cows, like all animals, are not themselves able to produce enzymes capable of digesting cellulose; so without help they can't extract the nutrients the grass contains. The cow overcomes this problem by having special fungi in the rumen called chytrids (more generally called rumen fungi; see Fig. 1.1).

These fungi are anaerobic, meaning they are able to survive without oxygen. Even without oxygen, chytrids are able to digest plant cell walls by making specific enzymes called cellulases. The rumen acts like a large fermenter because the grass is stored there while the fungal enzymes from the chytrids break down the cellulose and other components of the plant material to use as nutrients to produce more and more chytrids. This release of nutrients also supports the growth of large populations of bacteria and protozoa (which also have symbiotic associations with the chytrids and with one another) in the cow's gut. After the plant material is processed in the

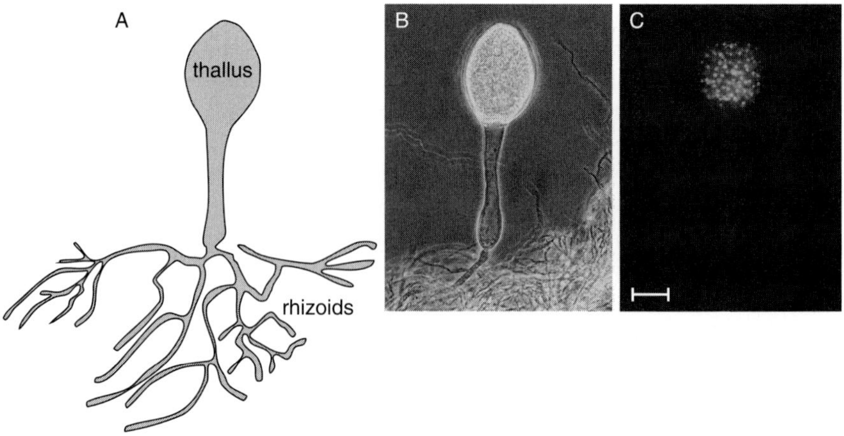

Fig. 1.1 General morphology of chytrids. **A**, sketch diagram of a chytrid thallus. The most obvious morphological feature is the thallus, the main body of the chytrid, perhaps 10 μm diameter, in which most of the cytoplasm resides and from which a system of branching elements emerges. The latter is called the rhizoidal system; the rhizoids anchor the fungus in its substratum and excrete digestive enzymes. The thallus is converted into a sporangium during reproduction, so the sporangium is sac-like and its protoplasm becomes internally divided to produce zoospores. **B**, phase contrast microphotograph of a thallus of the rumen chytrid, *Neocallimastix* sp., showing the single sporangium of the monocentric thallus and its rhizoidal system. **C**, the same field of view as in B, but this time with DAPI (4′-6-diamidino-2-phenylindole) fluorescence staining. DAPI forms fluorescent complexes with natural double-stranded DNA, so very specifically indicates nuclei when illuminated with UV light. In image C the fluorescent staining is limited to the thallus/sporangium, showing that the rhizoids do not contain nuclei. Scale bar = 40 μm. (Images modified from Moore, Robson & Trinci, 2011.)

rumen, it is brought back up into the mouth of the cow. This material is now called 'cud' and the cow chews it again to grind it down further. When it is swallowed for the second time it passes through the next three chambers of the stomach in which the large communities of microorganisms are digested by the cow (so the logic of the process is that almost indigestible plant material is turned into easily digested microorganisms). But this is not only of advantage to cattle, because this mutualism applies to sheep, deer, goats, and even llamas, giraffes and antelopes, and many more herbivores, large and small.

> The ruminant animal relies on microorganisms to convert the plant biomass into nutrients accessible to the animal. The concentrations of these populations in rumen fluid are in the range 10^9 to 10^{10} ml^{-1} bacteria; 10^5 to 10^6 ml^{-1} protozoa and about 10^9 ml^{-1} thallus-forming

> units of anaerobic fungi ... So an average cow contains about 1.5×10^{14} anaerobic fungi and an average sheep contains about 10^{11} anaerobic fungi.
>
> The total number of dairy cows in the world is 2.4×10^{8}, but in addition there are about 1.3×10^{9} beef cattle worldwide (source: FAO at http://www.fao.org/ag/aga/glipha/index.jsp). In total, therefore, there are approximately 1.6×10^{9} cattle worldwide, which amount to a grand total of 2.4×10^{11} litres of rumen fluid walking around containing 2.4×10^{23} anaerobic fungi in the world's cattle herd. Just the cattle herd. (Moore *et al.*, 2011, p. 494)

The chytrids are thought to pass from one animal to the next by being transferred in saliva, but they also occur in large numbers in cow dung. From the dung the fungi get attached onto surrounding grass. When another animal comes along and eats the grass, the fungi carry on their work in the new host. The relationship between chytrids and ruminants is mutualistic; in this case the animal benefits because plant material the animal can't degrade is digested and turned into materials (the microorganisms) the animal can digest and absorb. In return, the fungi live off some of the nutrients obtained from the animal's food, and live out their lives in the protected environment of the rumen. This relationship applies to all our farm animals, so it supports production of meat, such as beef and veal, as well as milk and dairy products, and other goods too, such as leather, wool and other fibres; each and every one depends on fungi in the animal's gut to digest the grass.

A consequence of that last statement is that our agricultural activities make indirect use of fungi. Our cereal and all other plant crops, from floribunda roses to forest trees, depend on mycorrhizal fungi; and all our farm animals depend on rumen fungi. Humans have been using fungi directly for many thousands of years. Many products that have been of great benefit to humans have been isolated from fungi or made with the help of fungi, especially food and beverages.

Beverages? Do you like a drink? Nice glass of wine, cool beer, soothing brandy? Every drop of alcoholic beverage production depends on the yeast fungus; and the annual value of this fermentation process today is in the region of $300 billion (and there's another $12 billion-worth of industrial alcohol produced every year). A little cheese goes well with a drink. Originally, milk was curdled with rennet (the enzyme chymosin) extracted from calves' stomachs, but about 90% of cheese today is made using chymosin from the mould fungus *Aspergillus niger*, or a fungal 'rennet' enzyme obtained from *Rhizomucor miehei*. The annual value of cheese production is about $35 billion, and every crumb depends on fungi. Like a little bread with your cheese? Every slice, or roll, or baton, or baguette of bread depends on fermentation by the yeast fungus to make the dough rise.

Would you like to add some sliced mushrooms to that ploughman's lunch? Fungi make a wholesome food, being nutritious with twice the protein of most fresh vegetables, a good source of fibre, and rich in minerals, essential amino acids and vitamins including B_2, niacin and B_{12}, few calories (36 in 100 g fresh weight) and little fat, but best of all, NO cholesterol. These advantages apply to yeasts, mycelium and mushrooms alike, but mushrooms offer a huge range of tastes and textures, which can be a delight for the adventurous cook. About ten species of mushroom are cultivated for sale in the United Kingdom; they contribute to one of the largest horticultural crops around the world, with a total annual value of about $50 billion.

Maybe you'd prefer a non-alcoholic beverage with your lunch; perhaps a can of cola? Ever looked at the ingredients list on your drinks can? A universally common ingredient is citric acid, which is widely used as an acidity regulator and antioxidant in foods and beverages and as a stabiliser in medicines. It was originally isolated from lemons, but this is expensive and this production method can't keep up with demand. But citric acid was found to be produced by *Aspergillus niger* fermentation in the 1920s and the process was industrialised in the 1940s. Today, about 600 000 tons of citric acid per year are produced by fermentation using *Aspergillus niger*, and every fizzy drink depends for its fizz on that fermentation.

As well as chemicals for food processing, fungi produce many digestive enzymes. In nature the fungi use these enzymes to digest organic material. That's their lifestyle. We use many of these enzymes in industry; that's our lifestyle. Just a few examples: fungal enzymes called pectinases increase fruit juice yields by 30% to 75%, pectinases are also used to 'peel' fruit and vegetables for canning and freezing; fungal enzymes are also used to make the food-sweetener fructose syrup; the current annual value of this industry is $15 billion.

And it's not just in food manufacture that fungi are crucial. Do you like stone-washed denim? Stone washing used to mean literally washing with small stones to 'distress' the fabric; but no more. Now fungal enzyme treatment randomly breaks cotton fibres giving the still-fashionable bleached look. We use similar fungal enzymes as 'fabric conditioners' to restore fabrics in the wash. In this case the enzyme removes broken fibre ends and makes the fabric look and feel like new. Biological washing powders also depend on fungal enzymes to remove soiling. Animal enzymes are no use because of the alkaline conditions produced by detergents. Proteinases (some bacterial in origin), lipases and amylases remove stains caused by proteins, fats and starches and other polysaccharides.

Some of the wonder drugs of today come from fungi. Most of us appreciate that if we have too much cholesterol the body is not able to use up the excess so it sticks to the inside walls of our blood vessels. This build-up reduces the diameter of the vessels, and this restricts blood flow. If blood vessels that supply blood to the heart become clogged up like this it can cause a heart attack, because the heart muscle does not receive enough oxygen to function properly. To control heart disease it's important that humans regulate their cholesterol level. The most effective cholesterol-lowering agents we have today are called statins, and these are produced by fungi. The two fungi used to produce statins are called *Aspergillus terreus* and *Penicillium citrinum*. Statins inhibit enzymes needed to make cholesterol, so production of cholesterol is slowed down. Add dietary control, and you can significantly decrease the patient's cholesterol level. Today, many people rely on statins from fungi to help keep their cholesterol level normal; the drugs are credited with saving 7000 lives a year in the United Kingdom alone.

Cyclosporine is another crucial wonder drug of today. It makes successful long-term transplant of livers, kidneys, hearts and lungs possible. This compound is produced by the fungus *Tolypocladium inflatum*. The fungus was isolated from a soil sample and screened in a search for antibiotics. The compound cyclosporine was found to be a weak antibiotic, but to have strong activity at suppressing the immune system (an immunosuppressive drug). This is the crucial role of this drug now. Our bodies are programmed to eliminate foreign things, and the body will naturally reject a transplant. The detection and elimination of foreign bodies is carried out by the immune system, which is made up of several cell types that act to protect our bodies from potentially harmful organisms. Lymphocytes are the cells that are able to detect foreign invaders. They attach themselves to pathogens identifying them as things to be destroyed. Following transplant operations cyclosporine helps prevent rejection by stopping the production of lymphocytes. If lymphocytes are not able to increase in number there is a greater chance that the transplant will not be detected and will continue to function normally. Cyclosporine is currently the most effective and widely used immunosuppressive drug. And we still depend on penicillin, the wonder drug of the 1940s, and on the ability of fungi to convert readily available plant steroids into compounds like contraceptives and anti-inflammatories. In this latter case the precursor is incubated with a live fungus culture and after a while the finished product can be isolated from the culture fluid. In a single process, the highly specific enzymes in the fungus make a chemical change that requires a 22-step organic synthesis for an industrial chemist to duplicate.

There is a medical downside because, although most of our ills are caused by viral and bacterial infections, fungi can infect plants and animals, including humans. Human fungal infections are divided into three groups. The first are superficial infections, which are infections of the outer layers of the skin, the hair and nails; examples are athlete's foot and ringworm (which is a fungus infection, not a worm). The second group are the subcutaneous fungal infections, where the deeper layers of the skin are infected, and sometimes even bone; the infective organisms usually cross the protective barrier of the skin at the site of a cut. Most of these organisms live in soil. Finally, fungal infections that enter into the body and invade internal organs are called systemic mycoses. Infection can arise from inhalation of fungal spores, although such cases are not usually life threatening. Most people that suffer from a systemic fungal infection are usually already sick with a condition that reduces the effectiveness of their immune system. If someone has such an immune deficiency their body is less able to defend itself against pathogenic organisms, and they therefore have an increased risk of susceptibility to infectious fungi. The fungus is said to be 'opportunistic' because if the person was healthy the fungus would not usually cause any serious harm.

The last few pages of this chapter have amply demonstrated that the rhythm of life on Earth in the present day benefits from several strong themes contributed by Kingdom Fungi. From the outline of contemporary fungal involvement in life on Earth given above it is abundantly clear that fungi make a crucial contribution to every ecosystem and an obvious conclusion is that this strong positive influence also existed in the past, and that the fungal lifestyle or body plan featured strongly during the evolution of life on Earth. If this is so clear, why has it not been remarked upon before now? Why are fungi ignored when theorists ponder the origin and early emergence of life on this planet? I am very sorry to say that I strongly suspect it is simply a display of ignorance.

In this chapter I have brought our focus to fungal organisms in an attempt to show the important roles that fungi play in life on Earth in the present day. If we are to explain how this set of circumstances arose through evolution we have to investigate in more detail those aspects of the cell biology of fungi that are unique to the fungi and that set them off as being so clearly different from animals and plants. This is done in the next chapter, but if you would like to investigate fungi further I suggest you refer to the book *Slayers, Saviors, Servants, and Sex: An Exposé of Kingdom Fungi* (Moore, 2000) for a general introduction to the activities of fungi and, particularly, the way humans interact with them; and/or for more comprehensive details, refer to the recently published textbook *21st Century Guidebook to Fungi* (Moore, Robson & Trinci, 2011).

TWO

ESSENTIALS OF FUNGAL CELL BIOLOGY

As I have shown in Chapter 1, with the quotation from Whittaker (1969), by about the middle of the twentieth century the three major kingdoms of eukaryotes were finally recognised, and a crucial character difference was their respective modes of nutrition:

(a) animals engulf
(b) plants photosynthesise
(c) fungi absorb externally digested nutrients.

As you might expect, many other differences can be added to these – some general differences, some highly specific. Some of these kingdom-specific differences are absolute, but most have to be qualified in some way. For example, you might, with some reason, say that a characteristic of animals is that they move, and contrast that with the characteristic immobility of plants. But coral reefs are made up of animals and yet are pretty immobile; and the next time you stroll through a meadow in late summer and the breeze stirs up an atmosphere filled with flying seeds, look around and remind yourself: 'the plants are migrating'.

Consequently, although it is possible to assemble panels of biological characteristics that are specifically expressed by each kingdom, you have to recognise that those characteristics may be subject to the context in which they are expressed and that in some circumstances there may be serious exceptions. When you try to establish evolutionary relationships there are more difficulties, the prime one being how to decide whether a character is ancestral or adapted. Intuitively, you might expect the ancestral character to be the simpler, and the adapted character to be the more complex. However, it's not always easy to recognise the difference; for example, the penguin and albatross must have had a common ancestor with wings, maybe as much as 50 to 100 million years ago; but which of those present-day birds has the most ancestral type of wing? Is it the one in which the wing is adapted to allow the animal to soar above the far ocean waves, or the one that allows the animal to swim beneath the same waves? Or might it be that they are both equally adapted, but in different behavioural directions? The issue must be decided by argument, and it's not so difficult to assemble that argument when you can venture out and observe the living organisms themselves, and observe other examples (petrels and guillemots make interesting comparisons with penguins and albatrosses), though the absolute relationships between these birds has not yet been 'decided'. By contrast, if your thoughts turn to more than a billion years ago, which is called 'deep' time, comparisons become much less easy. Unless the organism in which you are interested has become extinct, which has happened to a great many, then you still have their present-day descendants to study. But their deep time fossils may be rare and have, by definition, been subjected to billions of years of geological modification since the live organism was doing whatever it was doing. By this time, it may not be so easy to work out what it was doing.

With these warnings about complications for interpretation in mind I want next to examine which features of present-day fungi might characterise the kingdom in such a way as to give some indication of its evolutionary history. Fungi are often described as being characterised by one or two fungus-specific characters. For example, depending on context, it might be important to mention the following:

(a) In their cell (plasma) membranes animals use cholesterol as their major membrane lipid responsible for controlling membrane fluidity; in contrast, filamentous fungi use the chemically related compound ergosterol (these membranes separate the contents of the cell from the outside environment and provide the essential selective permeability that keeps the cell physiologically healthy; their chemical structure mirrors their specialisations).
(b) Recent genomic surveys show that plant genomes lack gene sequences that are crucial in animal development, and vice versa, and fungal genomes have none of the sequences that are important in controlling multicellular development in animals or plants, a feature that implies that animals, plants and fungi separated at a unicellular grade of organisation.
(c) Cell walls are located outside the cell membrane and their main function is to provide structural support, mainly to counteract the pressure of

water entering the cell (that is, crossing the membrane) by osmosis. Cell walls occur in bacteria, fungi, algae and plants; protozoa and animals do not have cell walls. In their cell walls, plants characteristically use cellulose (a glucose polymer), whereas fungi use chitin (a polymer made from *N*-acetyl-glucosamine).

(d) Filamentous fungi have an exclusively apical hyphal extension.

Without getting too involved, let's look at some of these features to see if any have sufficient promise to represent the fungi in deep time evolution. Cholesterol is, quantitatively, the predominant sterol in animal cell (or plasma) membranes. It serves to control membrane fluidity. Membrane fluidity determines the extent to which molecules can migrate through the membrane; fluidity is a function of the balance between fatty acid and cholesterol content. Fatty acid chains with unsaturated double bonds promote fluidity, whereas cholesterol molecules within the membrane interact with phospholipids to form lipid rafts that make the membranes less fluid. Changes in membrane-dependent functions, such as phagocytosis, the movement of molecules across the membrane, exposure of proteins at inner and/or outer membrane surface, and cell signalling, depend upon membrane fluidity, and cholesterol is the key variable in regulating this in animals.

Ergosterol probably fulfils a similar role in fungi; it gets its name from the fact that it was first isolated from the ergot fungus (*Claviceps purpurea*). Ergosterol is unique to fungi although both cholesterol and ergosterol are synthesised from a common precursor, lanosterol (Fig. 2.1). However, a very wide range of sterols has been detected in fungi and the spectrum of these can characterise strains. In addition, it is known that lipid, sterol and phospholipid contents differ between yeast and mycelial forms of *Candida albicans* (the yeast that causes skin diseases like thrush) and *Mucor lusitanicus* (a filamentous fungus like the bread mould), so there is a morphogenetic connection too. Most cells can synthesise a range of sterols, any of which might influence the physical properties of membranes and the activities of membrane-bound enzymes. In plants the most common sterols are stigmasterol, sitosterol and campesterol. Some fungi do not contain ergosterol, including chytrids (which use cholesterol) and the basidiomycete cereal rust-disease fungus *Puccinia graminis* (which uses fungisterol). Because ergosterol is unique to fungi it can be used to detect the presence of fungi. A strong correlation has been found between ergosterol content and fungal dry mass, even if the sample is contaminated with animal and/or plant tissues (for example, when measuring fungal growth in compost, soils or solid tissues, such as mycorrhizas on or in plant roots).

On first principles, the nature and structure of the cell wall looks like a fairly promising feature to make clear distinctions between groups of organisms; it's a tough, flexible to rigid coat that surrounds some types of cells.

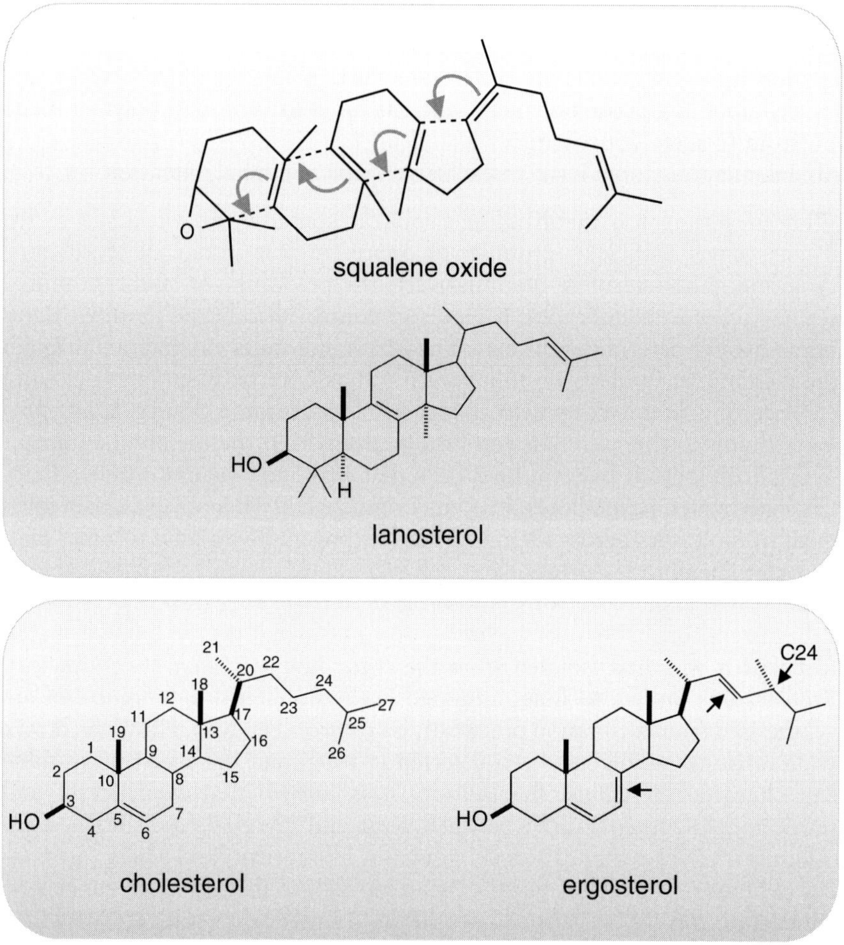

Fig. 2.1 Sterols are derived by cyclisation of squalene oxide as shown at the top of this illustration. Cholesterol and ergosterol are synthesised from a common precursor, lanosterol (centre). In the chemical structures of cholesterol and ergosterol (bottom) the cholesterol schematic shows the conventional numbering scheme for the carbon atoms and the ergosterol molecule shows how ergosterol differs from cholesterol by having an extra methyl group at C24 and two additional sites of unsaturation (arrows). (Modified from Moore, Robson & Trinci, 2011.)

Cells of higher animals and those of protozoa do not have cell walls, which distinguishes them from bacteria and other prokaryotes, algae, plants, fungi and fungus-like protists, which all have some form of cell wall. In all cases the wall provides the cell with protection and with structural support by containing the hydrostatic pressure generated when water enters the cell by osmosis; the wall, therefore, acts as a pressure vessel. The wall is, of course, external to the cell membrane, so it is constructed outside the cell and all maintenance and modification is done to it outside the cell. Consequently, having a cell wall implies that the cell also possesses a sophisticated mechanism for externalising all of the wall precursors, components, and synthetic enzymes and constructional aids that creation and maintenance of the wall may require. The materials used in construction of the cell wall differ between the different groups; the example is quoted above that the glucose polymer cellulose is the major structural component of cell walls in plants whereas fungi use a polymer made from *N*-acetyl-glucosamine (called chitin) for this purpose.

These examples show the common feature that polysaccharides (polymers of sugars) are common constituents of cell walls; the nature of the polymer depends on the sugar or sugars of which it is made, and there may be several different polysaccharides in any given wall. The polysaccharides are usually accompanied by various proteins (polymers of more than 50 amino acids) and peptides (which are short polymers of fewer than 50 amino acids) and the wall structure is varied by varying the way or ways in which these different components are chemically linked together. Algae typically have walls made of polysaccharides and glycoproteins, the latter being proteins that contain short chains of polysaccharide (also called oligosaccharide or glycan) chemically attached to the polypeptide. Probably the most unusual cell walls are those of diatoms, single-celled algae that produce intricately nanopatterned cell walls composed primarily of silica (SiO_2) with small amounts of polysaccharides (including chitin) and peptides.

Among the prokaryotes, peptidoglycans (also known as murein or mucopeptide) perform the essential physical support function in almost all bacteria (exceptions are wall-less mycoplasmas and archaebacteria) and are unique to bacteria; peptidoglycans are constructed of chains of polysaccharide (the glycan part of their name) which are cross-linked into two- and three-dimensional networks with short lengths of peptide (= small protein) chain, which contribute the 'peptido' part of the name. These structures are extremely important because these organisms are thought to be the most ancient, so the structure of their cell walls is likely to be primitive.

Archaebacterial cell walls have a variety of compositions, and may be made up of glycoproteins, peptidoglycans, and/or polysaccharides. This highly variable group in the present day contains many organisms that live in extreme environments (extremophiles) such as hot springs, alkali lakes, highly acidic rivers and salt flats.

Within any one of these groups, wall composition may differ between species, and within any one species can also differ depending on cell type and developmental stage. The most important example of a difference arising during development is that secondary walls in maturing plants

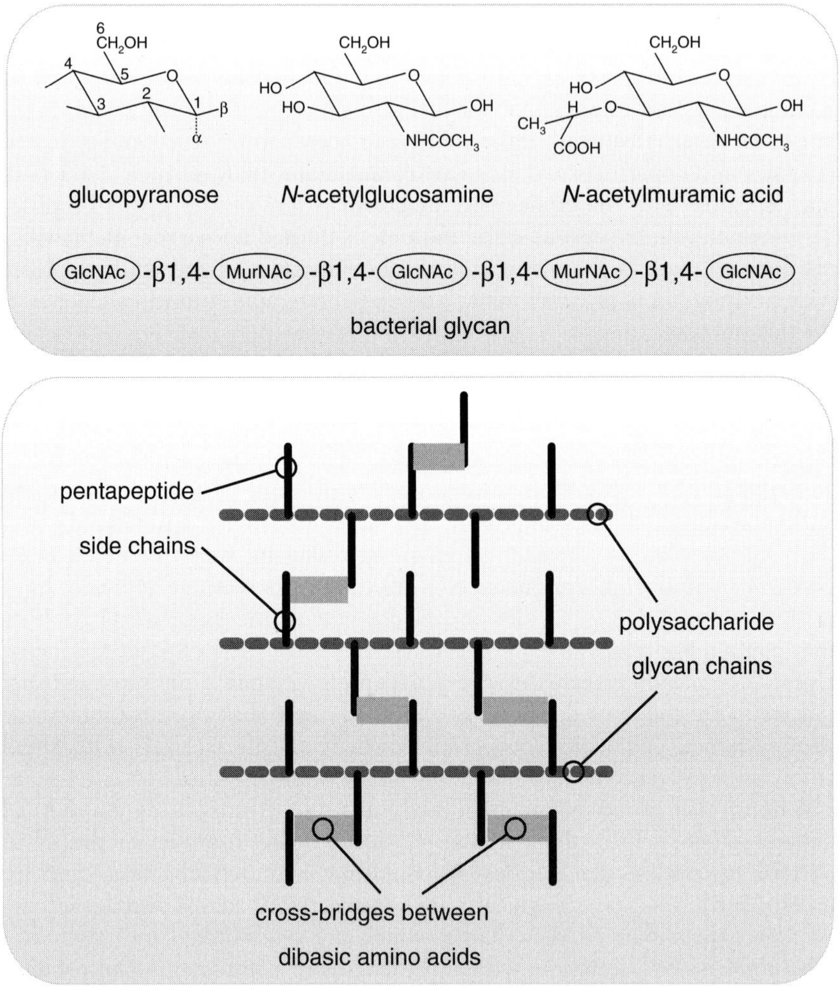

Fig. 2.2 General features of cell wall structure in bacteria, which is assumed to represent the primitive state. The upper panel shows the basic structure of sugar (glucopyranose) and molecular structures of *N*-acetyl-D-glucosamine (GlcNAc) and *N*-acetylmuramic acid (MurNAc). The glucopyranose molecule is a fairly flat six-membered ring composed of five carbons (numbered as shown) and one oxygen atom. The OH group on C1 (used to link sugars together in polymers) may be angled in the plane of the ring (β-orientation) or at right angles to it (α-orientation). In bacterial glycan, GlcNAc and MurNAc alternate in the polymer, being linked through β1,4 glycosidic bonds. The lower panel shows a cartoon drawing representing the completed peptidoglycan (murein) cell wall structure with the polysaccharide (glycan) chains being cross-linked by cross-bridges between short peptide side chains which are attached to the ether-linked (carbon-oxygen-carbon linkage) lactic acid

characteristically accumulate lignin. Lignin is a heterogeneous polymer of phenylpropanoid wood alcohols (*p*-coumaryl alcohol, coniferyl alcohol and sinapyl alcohol) that is highly hydrophobic, a feature that enables the plant vascular tissue to conduct water efficiently. The aromatic (or phenolic) nature of lignin also protects the plant cell wall against bacterial attack because casual degradation releases phenols that only fungi can degrade and which otherwise act like antiseptics.

On the basis that prokaryote (bacterial) cells were the first to evolve, the bacterial cell wall can be assumed to be the most ancient of the cell walls evident in present-day organisms. The characteristic peptidoglycans are based on a repeating unit of the amino sugars *N*-acetyl-D-glucosamine (GlcNAc) and *N*-acetylmuramic acid (MurNAc), with adjacent sugars linked β1,4 (Fig. 2.2). The glycan chains vary greatly in length; in *Staphylococcus aureus* the majority of chains have a length of 6 to 20 sugar units, with a maximum length in the region of 46 to 52 sugars, but length ranges of up to 200 sugar units in *Bacillus subtilis*, and 50 to over 160 sugar units in *Escherichia coli*.

The glycan chains are cross-linked via flexible peptide bridges to form a strong but elastic arrangement that contains the osmotic pressure of the bacterial protoplast. The chemical composition of the glycan chains varies only slightly, but the peptide bridges that are linked to the carboxyl group of MurNAc (see Fig. 2.2) are quite variable. They are synthesised as peptide chains of five amino acids, containing both L- and D-amino acids, and one dibasic amino acid, which contains two amino groups and allows the formation of cross-links between the peptide bridges (see Fig. 2.2). The dibasic amino acid is *meso*-diaminopimelic acid in most Gram-negative bacteria, and L-lysine in most Gram-positive bacteria. The most common stem

Caption for Fig. 2.2. continued

substituent ($CH_3CHCOOH$) of MurNAc in the glycan chains. The structures of the peptide side chains and cross-links are described in the text. **Definitions.** *Gram-positive or Gram-negative bacteria*: the amount and location of peptidoglycan in the prokaryotic cell wall is what determines whether a bacterium is Gram-positive or Gram-negative. Gram-positive bacteria have more layers of peptidoglycan in their cell walls than gram-negative, so they can retain the dye from Gram's staining method, which uses complex purple dye and iodine. *L- and D-amino acids*: all α-amino acids, except glycine, exist as either of two isomers, called L- or D-amino acids, which are mirror images of each other. L-amino acids are those found generally in proteins following ribosomal translation. D-amino acids are abundant components of the peptidoglycan in cell walls of bacteria, might also occur rarely in proteins as the result of post-translational modification of the L-isomer, and D-serine may act as a neurotransmitter in the brain.

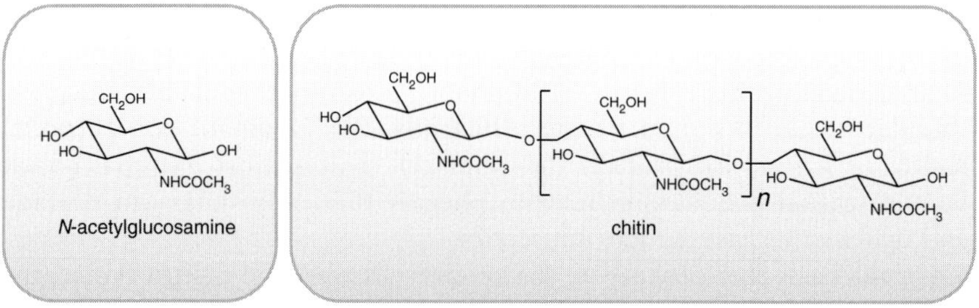

Fig. 2.3 Structural formulae of *N*-acetylglucosamine and its linear homopolymer, chitin, which is synthesised by the enzyme chitin synthase. Natural sources of chitin have molecular masses of a few million; extraction processes fragment the polymer and the molecular mass of commercial preparations of chitin can vary between 350 000 and 650 000; these molecular masses correspond to polymers made up of 2000 to 20 000 glucosamine units (compare with the 100 to 200 sugar units that are characteristic of bacterial glycans). Chitin structure is similar to that of cellulose (Fig. 2.6), except that acetylamino groups replace the hydroxyl groups at carbon position 2 of the glucose molecule. Approximately 16% of naturally occurring chitin units have the acetyl group removed after assembly of the polymer. (Modified from Moore, Robson & Trinci, 2011.)

peptide in *E. coli* and *B. subtilis* is L-alanine (1), –D-glutamate (2), –*meso*-diaminopimelic acid (3), –D-alanine (4), and –D-alanine (5); with the L-alanine (number 1 in the previous list) being attached to the carboxyl group of MurNAc (Fig. 2.2). The cross-links between peptides are formed by a transpeptidase enzyme linking D-alanine (number 4) from one peptide to the free amino group of *meso*-diaminopimelic acid (number 3) on another peptide (Fig. 2.2); in *S. aureus* this cross-linking is made through an amino acid bridge made up of five glycine molecules. The proportion of glycan chains cross-linked varies: 40% of total chains in *E. coli* and 100% of total chains in *S. aureus*. Each chain may be linked to two others and up to ten glycan chains may be attached through the cross-linkages.

I am stressing these details because they will become relevant later in this discussion. Later we will be deriving eukaryotes from a bacterial (specifically, actinomycete) ancestor and I want you to note and remember the similarity between the characteristic cell wall glycan of bacteria [GlcNAc + MurNAc + GlcNAc + MurNAc ...] and the characteristic cell wall chitin of fungi [GlcNAc + GlcNAc + GlcNAc + GlcNAc ...], as there seems to be scope here for adapting the synthesis of bacterial glycan to the production of fungal chitin.

Most true fungi have a cell wall consisting largely of chitin (Fig. 2.3) and other polysaccharides arranged as a multilayered complex of polysaccharides, glycoproteins and proteins. Generally speaking, there are usually three layers of cell wall material (Fig. 2.4):

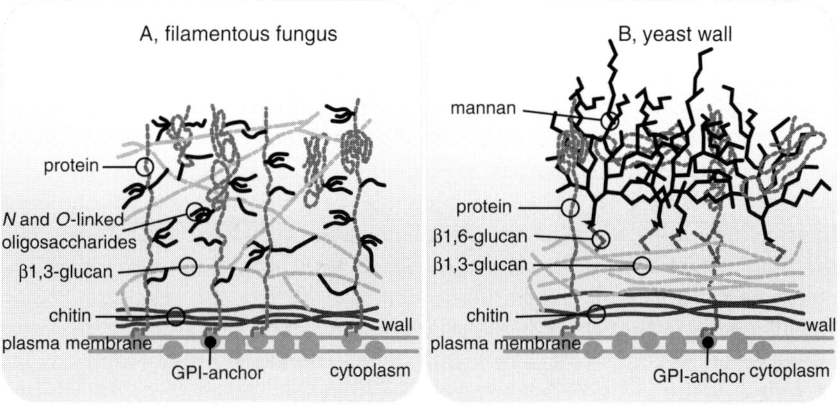

Fig. 2.4 Diagrammatic depictions of the structure of fungal walls. **A**, interpretation of the filamentous hyphal cell wall. Most of the chitin is thought to be near to the plasma membrane where it resists the turgor pressure of the protoplast; β1,3-glucan extends throughout the wall. Protein, glucan and chitin components are integrated into the wall by cross-linking them together with *N*- and *O*-linked oligosaccharides. Many of the glycoproteins have GPI anchors (*G*lycosyl*P*hosphatidyl*I*nositol anchors), which link the protein to the plasma membrane, while other glycoproteins are secreted into the wall matrix. **B**, generalised cell wall structure of the yeast *Candida albicans*; note that mannans (shown in black) dominate the outer regions of this yeast wall and β1,6-glucan cross-links the components. Also note that in liquid cultures both types of wall 'blend into' the surrounding aqueous medium because the surface polymers (polypeptide and polysaccharide) are hydrophilic and able to dissolve into the aqueous medium. Walls of aerial hyphae can be chemically modified at the surface by deposition of polyphenols and/or assembly of layers of hydrophobic proteins. Some GPI-anchored wall proteins span the wall with extended glycosylated polypeptides protruding into the surrounding medium and providing the wall with hydrophilic and even adhesive surfaces. (Modified from Moore, Robson & Trinci, 2011.)

(a) Closest to the plasma membrane there is a chitin layer consisting of unbranched chains of *N*-acetyl-D-glucosamine;
(b) then a layer of β1,3-glucan (a glucan is a polysaccharide made mostly from glucose); and
(c) on the outside, a layer of mannoproteins (glycoproteins in which the sugar appendages are mainly mannose).

Other complex polysaccharides, such as glucogalactomannans (polysaccharides made up of mixtures of the sugars glucose, galactose and mannose) may be scattered through the wall.

In hyphae the chitin is the major component of the wall; walls of hyphae of *Neurospora* and *Aspergillus* contain 10–20% chitin, and although the chitin molecules are frequently cross-linked to other wall constituents, particularly the β1,3-glucan, it is the chitin component that provides the main structural strength of the wall. Chitin accounts for only 1–2% of the yeast cell wall by dry weight, although yeast chitin is mainly found in bud scars where it provides a strong seal when the bud separates from the mother cell. In both types of fungal cell, though, chitin molecules form into microfibrils, about 10 nm in diameter, by head-to-tail 'crystallisation' between the polymers. These crystalline polymers are held together by spontaneous hydrogen bonding and the microfibrils have the tensile strength to provide the wall with its main structural integrity. In yeasts, the major structural component of the wall is a fibrillar inner layer of β-glucan (Fig. 2.4).

The glucan in fungal walls has secreted mannoproteins attached to it and these play an essential role in cell wall organisation. The protein components of fungal walls are very important; they may be anchored in the plasma membrane, covalently bonded to wall polysaccharides or more loosely associated with the wall. Some of the wall proteins are enzymic, other proteins are involved in cell-to-cell recognition and/or cell-to-cell adhesion. The outer surface of many fungal walls is usually layered with proteins that modify the biophysical properties of the wall surface appropriately to the environment. The synthesis and assembly of fungal walls is worth illustrating at this point (Fig. 2.5) because the diagram emphasises that wall synthesis involves the activity of the main membrane-bound compartments of the cell (particularly the endoplasmic reticulum and the Golgi compartments); the production of a great number of vacuoles and vesicles; and the (rapid) transport of these components to the apex of the hypha where wall synthesis in a filamentous fungus is concentrated. In the present day, everything happens quickly. It has been calculated that 38 000 vesicles have to fuse with the apical membrane each minute (that's over 600 every second) to support extension of each hyphal tip of *Neurospora crassa* when it is growing at its maximum rate (Moore *et al.*, 2011, see sections 5.12 and 5.15, and chapter 6). The management of vesicle production, transport and delivery to these levels of speed and precision is a major evolutionary achievement for the fungi.

In the plant cell wall cellulose predominates; indeed, in the present day cellulose is the most abundant organic compound on Earth and accounts for over 50% of organic carbon on the planet; about 10^{11} tonnes are synthesised each year. Cellulose is an unbranched polymer of glucose in which adjacent sugar molecules are joined by β1,4-linkages (Fig. 2.6); there may be from a few hundred to a few thousand sugar residues in the polymer molecule, corresponding to molecular masses from about 50 000 to approaching 1 million. Comparison with chitin (compare Figs. 2.3 and 2.6) shows that the difference between the two polymers is the presence of the acetylamino ($-NHCOCH_3$) group on each and every sugar molecule of which the chitin molecule is made; it follows that the larger chitin molecules to which reference is made in the legend to Fig. 2.3 that contain 20 000 glucosamine units

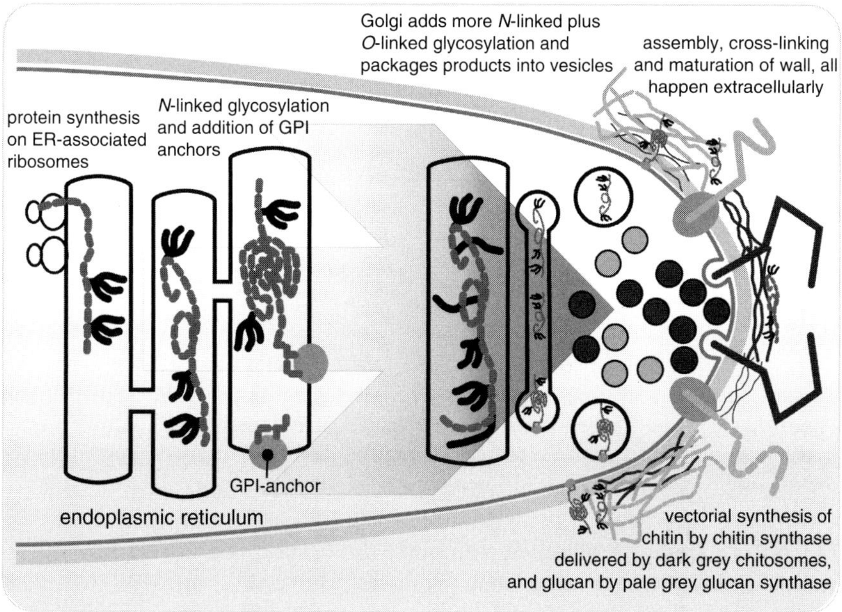

Fig. 2.5 Biosynthesis of cell wall components. Glycoprotein synthesis begins in the endoplasmic reticulum (ER) with the addition of *N*-linked oligosaccharides (black) during translation. GPI anchors are also added to some proteins in the ER. In the Golgi apparatus, the glycosyltransferases modify the proteins further by addition of sugars to generate *O*-linked oligosaccharides and to extend *N*-linked oligosaccharides. Glycoproteins are secreted into the cell wall space where they are integrated into the cell wall structure. The chitin (dark grey) and glucan (light grey) components of the cell wall are vectorially synthesised on the plasma membrane and extruded into the cell wall space during their synthesis. The various components of the wall are cross-linked together in the cell wall space by cell-wall-associated glycosylhydrolases and glycosyltransferases. (Modified from Moore, Robson & Trinci, 2011.)

require 20 000 atoms of reduced nitrogen. This is nitrogen which is 'frozen' into the cell wall; unless the wall is recycled (a very drastic prospect) those nitrogen molecules cannot be recycled into the general metabolism of the cell. So wall synthesis in fungi costs a lot of nitrogen. For a plant cell, the molecules of cellulose the cell makes for its cell wall, no matter how large they may be, contain absolutely no nitrogen. This difference offers a selective advantage in favour of the use of cellulose by plant cell walls. Let me explain why.

The 'nitrogen cost' of their wall is something the fungi can afford because they are heterotrophs. When proteins and other nitrogen-containing compounds are broken down, either as part of the turnover process within the

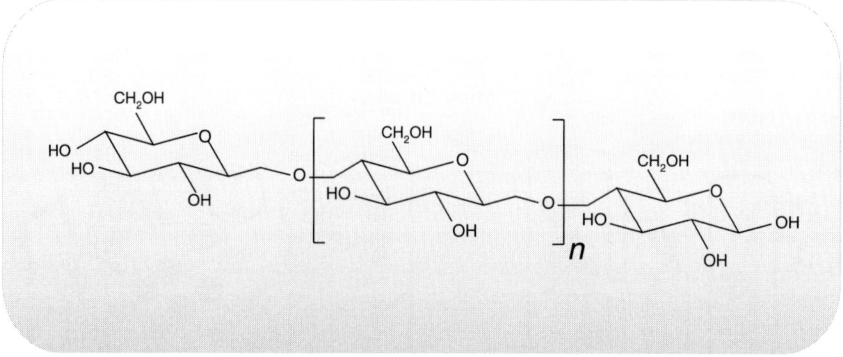

Fig. 2.6 Structural formula of cellulose. There may be from a few hundred to a few thousand sugar residues in the polymer molecule, corresponding to molecular masses from 50 000 (300 sugar units) to approaching 1 million (about 6000 sugar units in the polymer). (Modified from Moore, Robson & Trinci, 2011.)

cell or as externally obtained nutrients, the carbon is recovered to use in energy metabolism. Using protein produces an excess of nitrogen that must be excreted. Experiments with some common mushrooms, *Agaricus bisporus* (the cultivated mushroom), *Coprinopsis cinerea* (a common ink cap field mushroom) and *Volvariella volvacea* (the paddy-straw mushroom) have shown that one third to one half of the nitrogen contained in protein given as substrate is excreted as ammonium into the medium. This is similar to terrestrial mammals that metabolise protein and excrete the excess nitrogen as urea formed through the urea cycle. Plants, though, do not metabolise external supplies of protein; their basic nutritional mode (photosynthesis) involves only water (H_2O) and carbon dioxide (CO_2). Nitrogen, which the plant needs to make all of its proteins and nucleic acids, can only come directly to the root from the soil water as nitrate (NO_3) or ammonium (NH_4). Later in this book I will argue that the photosynthetic plants diverged from the heterotrophic primitive eukaryotes and it seems to me reasonable to expect that those early plants will have found selective advantage in adapting their cell walls to use cellulose rather than chitin in order to economise on their demand for nitrogen.

I have so far discussed the specific characteristics of fungi that are most often acknowledged; I might even suggest that in certain quarters these are the only ones known. But, like the other eukaryote kingdoms, fungi are characterised by an integrated suite of unique cell biological features that contribute to what might be called their body plan (though I prefer the word lifestyle). I have already indicated above that fungal wall synthesis involves the activity of the main membrane-bound compartments of the cell and the rapid transport of a great number of vacuoles and vesicles to the apex of the hypha (summarised in Fig. 2.5); stating that (and it merits repetition) 'The management of vesicle production, transport and delivery to these levels of speed and precision is a major evolutionary achievement for the fungi'. For filamentous fungi, which are in the majority in the present day, the culmination of that evolutionary achievement is a complex organelle which is absolutely unique to fungi. It is called the Spitzenkörper, which is German for 'apical body' because it can be detected using conventional light microscopy in the extreme apex of growing hyphae of ascomycetes and basidiomycetes (the two groups of 'higher' fungi; the 'moulds' and the 'mushrooms'). Some of the more primitive zygomycetes do lack a characteristic Spitzenkörper, but instead have a loose accumulation of vesicles in the hyphal apex that may serve the same function. The Spitzenkörper is present in actively growing hyphal tips but is lost when apex extension stops, so it is clearly the organising centre for hyphal extension and morphogenesis.

The Spitzenkörper is clearly a complex organelle, which is composed of vesicles of various kinds and sizes, cytoskeletal microfilaments and microtubules, and active ribosomes and may interact with nearby mitochondria. It is not yet fully understood but because the Spitzenkörper is always adjacent to the site of polarised hyphal extension, and because extension of the hyphal tip requires polarised incorporation of plasma membrane and cell wall constituents into the growing apex, the best interpretation seems to be that the Spitzenkörper serves as a 'vesicle supply centre'. On this interpretation endoplasmic reticulum and Golgi tubules and cisternae further back in the body of the hypha generate secretory vesicles containing wall and membrane components and ship them forwards using molecular motors on the cytoskeletal network. These are collected into the Spitzenkörper, matured, coordinated, and then discharged forwards to extend the hyphal apex. The discharge emerges from the Spitzenkörper like the light from a spotlight; it always shines forwards, but it can be steered. If the Spitzenkörper is steered 15° to the right, then the hyphal apex will be directed to the right and will describe a rightward curve. When that Spitzenkörper is returned to the 0° angular position the turn will end and the hypha will extend forwards on its new track. By this manoeuvre the hyphal tip may have avoided an obstacle in its original path or been directed towards a fresh supply of nutrients. This little story makes the point that the behaviour of the hypha (each and every hypha) depends on its Spitzenkörper and the information its Spitzenkörper receives and acts upon; that is, as well as discharging vesicles forwards to extend the hyphal apex, the Spitzenkörper must have the ability to receive, interpret, and act upon signals received from outside the hypha.

The Spitzenkörper must rank as one of the most sophisticated cytoskeletal organising centres in the whole of biology. It is effectively 'spraying' vesicles and microvesicles forwards to the apex; but it can't work like a paint sprayer using air pressure. Instead it must create filamentous cytoskeletal tracks (microfilaments, intermediate filaments, and microtubules), each one targeted to a specific position in three-dimensional space, and then load those tracks with molecular motors laden with the cargo required at the destination and provide those motors with the energy supplies that will enable them to arrive at the specific time to permit a seamless apical synthesis. And in *Neurospora crassa*, each Spitzenkörper is potentially doing all this for over 600 vesicles every second in each hyphal tip; and each still has time to receive, interpret, and act upon the latest signals from the outside.

This pinnacle of cytoskeletal management could not have been achieved with a few mutational adaptations, so the question I need to address is whether there is any evidence in present-day fungi of characters that might have contributed to a pathway of evolutionary change in the past that could culminate with the Spitzenkörper. I think the answer rests with that integrated suite of unique cell biological features that make up the characteristic fungal body plan/lifestyle. These are features which are unique to the fungi, and are well known, but are rarely if ever brought together in this organised way. Specifically, these are:

(a) free cell formation, the process that 'invents' a cytoskeletal management system for organelle location and vesicle distribution to subdivide volumes of cytoplasm to make spores;
(b) filamentous growth, with the Spitzenkörper as the ultimate organising centre that makes nucleated exploratory hyphae, the process that invents the filamentous fungal lifestyle;
(c) hyphal/cell fusion, the process that enables the invention of individuality and cell-to-cell interactions such as sexual reproduction;
(d) hyphal-septum formation, the process that invents multicellularity.

These are the characters that must be arranged in the temporal order shown above to explain the phylogenetic origin of eukaryotes in general and the fungi in particular (discussed in Chapter 12). All of these features rank as revolutions in cell structure (using 'revolution' in the sense of Cavalier-Smith (2006) as adaptations capable of precipitating major evolutionary advances), though they are seldom viewed as being so important.

Free cell formation is the expression of a highly organised cytoskeletal system that manages organelle positioning and the distribution of wall- and membrane-forming vesicles to enclose volumes of cytoplasm to subdivide sporangia into spores. It is a unique feature of fungal biology that was first described thoroughly following use of the electron microscope to examine the ultrastructure of zoospore formation in the chytrid *Blastocladiella* (Lessie & Lovett, 1968). Its significance has been illustrated this way:

> ... let's carry out the thought experiment of working out what would happen if these fungi were either animals or plants. The situation is that

> we are converting the chytrid thallus, a single sac-like cell, into a sporangium. Initially there is a single nucleus, but this will undergo several mitotic divisions so that the volume of the sporangium can be subdivided into many zoospores, each of which will have a single mitotically produced nucleus. How will that subdivision be managed?
>
> If *Blastocladiella* was an animal then at each division the dividing cell would become constricted at the equator of the mitotic spindle and two daughter cells would be produced as a result of the cleavage of the mother cell. Through successive rounds of mitosis, more and more cells would be produced; just like a developing animal embryo.
>
> If *Blastocladiella* was a plant then at each nuclear division a daughter cell wall would be formed across the equator of the mitotic division spindle. Daughter cells would then be successively halved in size (but doubled in number) as each round of mitosis occurred.
>
> But *Blastocladiella* is neither animal nor plant, and it does neither of these things. Instead, *Blastocladiella* uses a uniquely fungal mechanism. Its zoospores are formed by cleavage of the multinucleate protoplasm in the zoosporangium, yes, but this happens as masses of cytoplasmic vesicles fuse to one another to create the borders between adjacent zoospores. (Moore *et al.*, 2011, p. 48, with more discussion and illustration on subsequent pages)

This remarkably precise zoospore-generating pattern is repeated throughout the chytrids, and indeed this mechanism for subdividing cytoplasm, which depends on the organised distribution of cytoplasmic vesicles (Fig. 2.7), pervades the entire fungal kingdom. Compare Fig. 2.7 with this description of sporogenesis in the mucoraceous (terrestrial) fungus *Gilbertella persicaria*:

> During cleavage, the principal structural changes involve pattern transformations of protoplasmic membranes ... small vesicles are formed, apparently from special cisternae [of the endoplasmic reticulum]. The disappearance of these initial vesicles coincides with the appearance of cleavage vesicles ... distinguished by the presence of granules on the inner surface of the vesicle membrane ... Cleavage is initiated endogenously by the coalescence of cleavage vesicles to form a ramifying tubular cleavage apparatus. The cleavage apparatus demarcates the boundaries of potential spore initials. Lateral expansion of elements of the cleavage apparatus results in furrow-like configurations which converge to cut out spore initials as independent cells. The cleavage membrane is transformed to the plasma membrane of spore initials during late cleavage ... The marker granules that were present around the periphery of the cleavage vesicles are found on the outer surfaces of spore plasma membranes after cleavage. The granules fuse to form a continuous spore envelope, and subsequently the spore wall is laid down centripetally. Thus, the envelope becomes the outermost spore wall layer ... (Bracker, 1968)

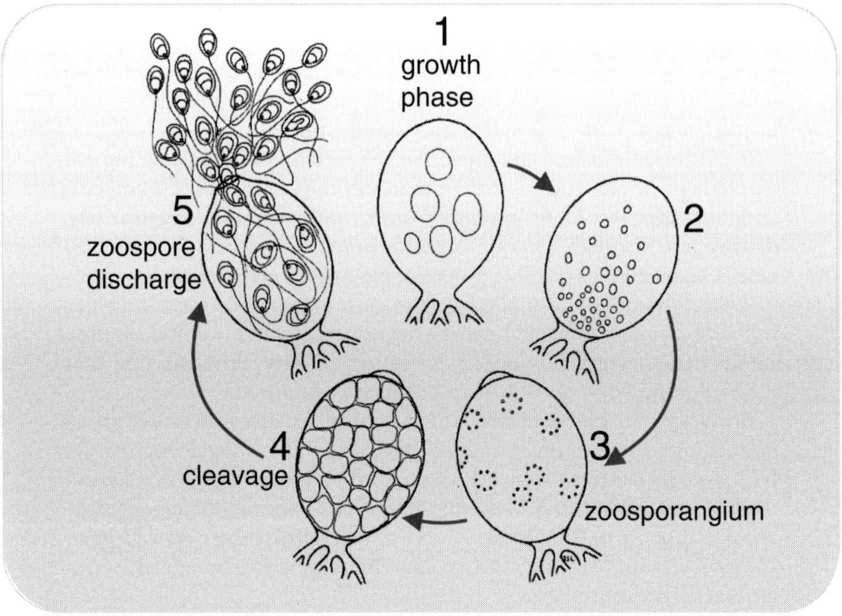

Fig. 2.7 The process of sporangium formation and zoospore differentiation in *Blastocladiella*. The diagrams illustrate the changing appearance of sporangia during differentiation, as seen in the light microscope. The original authors describe the process as follows: 'Soon after the beginning of flagella formation it is possible to find early stages of "cleavage furrow" formation . . . This process . . . involves the fusion of many small vesicles . . . cleavage vesicle fusion results in progressive expansion of the primary cleavage furrows and it appears that this activity is simultaneously initiated at many points. Occasionally vesicles can be found in somewhat linear arrangements over a short distance. They more often occur in less orderly clusters and fuse in irregular ring-shaped patterns lying roughly in the plane of the developing cleavage furrow. The frequent occurrence of cytoplasmic peninsulas surrounded by U-shaped areas of cleavage vesicle suggests that many of the rings may in fact be short cylinders; if so, the closure and interconnection of the rings may be irregular and only gradually assume the form of a regular furrow . . . The cleavage furrows also fuse with the earlier formed vesicles surrounding the flagella with the result that these finally lie within the cleavage furrows and outside of the uninucleate blocks of cytoplasm delineated by the [newly formed] membrane system' (Lessie & Lovett, 1968). (Modified from the original drawing by James S. Lovett in Lessie & Lovett, 1968, which illustrates all of these features with electronmicrographs, though it does refer to the *Blastocladiella* thalli as both 'fungus' and 'plant', something which would not be done now. Reproduced with permission of the Botanical Society of America.)

The process described here has been called 'free cell formation' and has been shown to apply to ascospore formation in ascomycetes, and to the early stages of basidiospore development in basidiomycetes, though in the latter it is heavily modified as nuclei are forced to migrate into spores by a vacuolation process (Moore *et al.*, 2011, p. 50). These authors emphasise the general rule that where a volume of cytoplasm needs to be subdivided in fungi, the mechanism depends on the organised distribution of cytoplasmic microvesicles; the microvesicles then fuse together to create the separation of the cytoplasm. This is the way the fungi do it (and a similar cleavage system produces zoospores in sporangia of the fungus-like Oomycota), but this is a major difference from plants (walls formed across the equator of the division spindle) and animals (which use constrictive cell cleavage across the equator of the division spindle). Significantly, free cell formation is characteristic of the most primitive true fungi, the chytrids; so this can be categorised as the first step towards cytoskeletal management. A step taken at the earliest stage in eukaryote evolution that has persisted into modern fungi, but one discarded by plants and animals as they diverged from the eukaryote stem, though in plants remnants of free cell formation may have been adapted into the phragmoplast that characterises plant cell division.

Apically extending filamentous growth, with the Spitzenkörper as the organising centre that collects and distributes wall- and membrane-forming cytoplasmic vesicles, is often cited as a unique fungal characteristic. I have given some description above and it is described in detail by Moore *et al.* (2011, pp. 137–142). The Spitzenkörper is thought to be a vesicle supply centre (VSC). It receives vesicles from the endomembrane system (Golgi and endoplasmic reticulum) behind the tip and then wall-building vesicles are distributed from the VSC. Vesicles move radially to the hyphal surface in front of the VSC. Fusion of vesicles with the plasma membrane externalises their content of substrates and lytic enzymes (endoglucanase, perhaps chitinase) and hydrolyses structural glucan molecules in the existing wall so that as mechanical stretching pulls the broken molecules apart resynthesis occurs either by insertion of oligoglucan or by synthetic extension of the divided molecule(s). The resynthesised molecules have the same mechanical strength as before, but have been lengthened and the tip has grown. Forward migration of the VSC generates the shape of the hyphal tip (Riquelme & Bartnicki-García, 2008; Riquelme *et al.*, 2007; Steinberg, 2007).

During early growth of a mycelium, nutrients surrounding the young mycelium are in excess and the mycelium is unrestricted and undifferentiated. During undifferentiated growth, the mean rate of hyphal extension is dependent on the specific growth rate of the organism (the maximum rate of growth in biomass per unit time) and the manner and degree of branching. In older mycelium, hyphal fusions are evident at the colony centre, as are hyphal

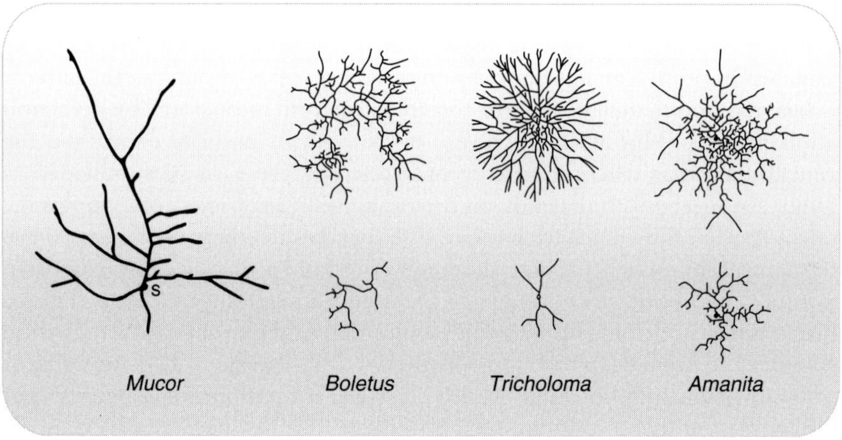

Fig. 2.8 These hand-drawn pictures illustrate the well-ordered, young mycelium which colonises solid substrates effectively and efficiently. From the left: a young colony of *Mucor* (S, position of the spore from which the rest of the hyphae have germinated); the rest show a very young germling below and a slightly more mature colony above, of *Boletus*, *Tricholoma* and *Amanita*. (Modified from Moore, Robson & Trinci, 2011)

avoidance reactions at the colony margin. Between them, the sketches in Figs. 2.8 and 2.9 portray the main growth processes that influence the distribution of hyphae in a mycelium (which constitute the essential fungal lifestyle), which are:

(a) polarised hyphal growth;
(b) branching frequency;
(c) autotropism (the 'self-avoidance' reaction that makes vegetative hyphae grow away from the already-existing mycelium).

It is important to recognise that these are dynamic relationships. They change with the age and with the developmental state of the mycelium as its biological functions change; indeed the distribution of biomass in a fungal colony varies with the age of the hyphae. One part of a mycelium may be growing as a rapidly extending, sparsely branched exploratory sector, another part may be a highly branched and interconnected network exploiting a nutrient resource, while a third region reverses the autotropism so that hyphal tips congregate and co-operate to form a fruiting or survival structure like a sclerotium or mushroom.

In mycelial fungi of the present day, hyphal fusions are an aspect of mycelial maturation; lateral branches and hyphal fusions convert the central regions of a maturing colony into a fully interconnected network through which materials and signals can be communicated efficiently and this enables

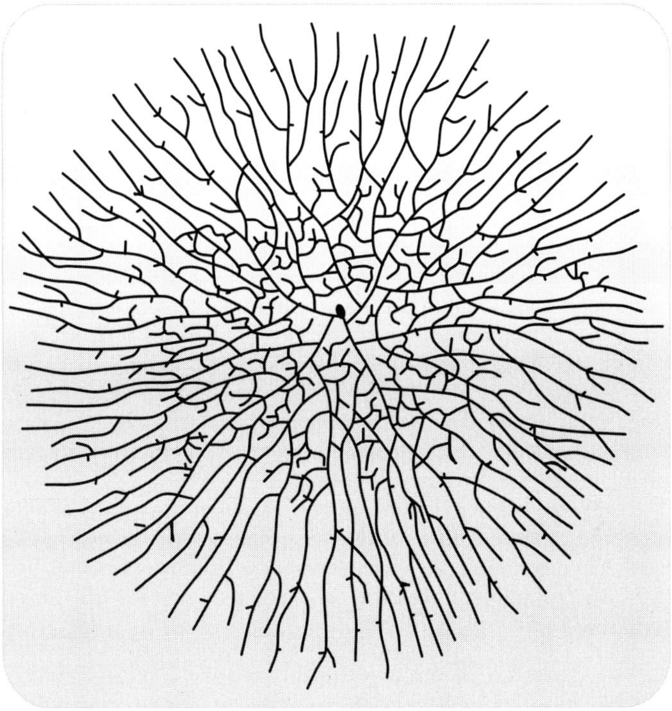

Fig. 2.9 A maturing fungal colony. Notice how the growing hyphae are oriented outward into uncolonised regions (negative autotropism) whilst the production of branches and hyphal fusions centrally ensures the mycelium becomes a network that efficiently exploits available substrate (and these branches show positive autotropism because they grow towards their sister hyphae). (This hand-drawn sketch of *Coprinus sterquilinus* comes from vol. 4 of A. H. R. Buller's epic series *Researches on Fungi*. Modified from Moore, Robson & Trinci, 2011.)

the vegetative mycelium to make best use of its resources. A hyphal tip approaching another hypha may induce a branch to which it subsequently fuses; alternatively, two hyphal tips may approach each other and then fuse; or a hyphal tip approaches the side of another existing hypha and there is a tip-to-side fusion:

> The Spitzenkörper is closely involved in the processes of hyphal fusion, which have been characterised into nine stages (Glass *et al.*, 2004) as follows:
>
> Stage 1: a hyphal tip competent to initiate fusion secretes an unknown diffusible, extracellular signal which induces Spitzenkörper formation in the hypha it is approaching; it is presumed that the second hypha secretes a corresponding signal;

> Stages 2 and 3: these events result in two fusion-competent hyphal tips each secreting diffusible, extracellular chemotropic signals that regulate Spitzenkörper behaviour so that the hyphal tips grow towards each other . . .;
>
> Stage 4: cell walls of the approaching hyphal tips make contact, apical extension ceases, but both Spitzenkörpers persist;
>
> Stage 5: adhesive material is secreted at the hyphal tips;
>
> Stage 6: polarised apical extension is converted to 'all-over' isotropic growth, which results in swelling of the adherent hyphal tips;
>
> Stage 7: cell walls and adhesive material at the point of contact are dissolved, bringing the two plasma membranes into contact;
>
> Stage 8: plasma membranes of the two hyphal tips fuse, creating a pore with which the Spitzenkörper stays associated as the pore begins to widen and cytoplasm starts to flow between the now connected hyphae;
>
> Stage 9: the pore widens, Spitzenkörper disappears, organelles, including nuclei, vacuoles and mitochondria, can flow between the fused hyphae, though the flow may be regulated.
>
> The principal lesson to take away from this brief description of hyphal fusion can be phrased like this: what the Spitzenkörper has put together, the Spitzenkörper can take apart. (quotation from Moore *et al.*, 2011, pp. 142–144; and see Read *et al.*, 2009, 2010 for the original descriptions)

Enhancing the efficiency of the mycelium by converting an initially radially spreading collection of hyphae into an integrated network would be the greatest adaptive value of hyphal fusions. In present-day fungi, which make a variety of multicellular tissues within their fruit bodies and resting structures (such as sclerotia) differentiated to specific functions,

> ... lateral contacts between fungal hyphae are extremely rare, being represented only by lateral hyphal fusions. The constituent cells of plant and animal tissues are interconnected laterally by frequent plasmodesmata, gap junctions and cell processes. The absence of similar structures connecting adjacent hyphae suggests that any morphogens which do exist are likely to be communicated exclusively through the extracellular environment ... (Moore, 1998, p. 286)

Once the process of hyphal fusion is established it can be adapted to interactions between different individual mycelia. This would enable exchange of nuclei: first to form the heterokaryons that are so important in the biology and evolution of present-day fungal populations (Taylor, Jacobson & Fisher, 1999), and, second, as a prelude to sexual reproduction and all that may mean for evolutionary progress.

In the present day the hyphal growth form of filamentous fungi is an adaptation to the active colonisation of solid substrata. Exploration by extending hyphal filaments into the environment is hazardous for hyphae that lack cross walls; any puncture to the osmotically pressurised hydrostatic system can lead to uncontrollable loss of cytoplasmic contents. There is, consequently, selective advantage in the ability to make cross walls (septa). In current fungi:

> The septa which divide hyphae into cells may be complete (imperforate), penetrated by cytoplasmic strands, or perforated by a large central pore. The pore may be open (and offer little hindrance to the passage of cytoplasmic organelles and nuclei), or may be protected by a complex cap structure derived from the endoplasmic reticulum (the dolipore septum of Basidiomycota). In Ascomycota, the pore may be associated with, perhaps plugged by, Woronin bodies, which are modified peroxisomes. Septal form may be modified by the hyphal cells on either side of the septum, and may vary according to age, position in the mycelium, or position in the tissues of a differentiated structure. These features make it clear that the movement or migration of cytoplasmic components between neighbouring compartments is under very effective control. The hypha is divided up by the septa, and the cellular structure of the hypha extends, at least, to its being separated into compartments whose interactions are carefully regulated and which can exhibit contrasting patterns of differentiation ... Filamentous fungi form multinucleate hyphae that are eventually partitioned by septa into multicellular hyphae. In *Aspergillus nidulans*, septum formation follows the completion of mitosis and requires the assembly of a septal band ... This band is a dynamic structure composed of actin, septin and formin. Assembly depends on a conserved protein kinase cascade that, in yeast, regulates mitotic exit and septation. (Moore *et al.*, 2011, pp. 144–150; and see Steinberg & Schuster, 2011, for illustration of the dynamic behaviour of major cytoskeletal elements and organelles in fungal cells)

The mechanical geometry of the original (ancient) septal band, operating as a cytoplasmic sphincter if the wall is punctured, presumably accounts for that other feature of fungal cross walls, which is characteristic of fungi, namely that septa are always placed at right angles to the long axis of the hypha, or segment of hypha, within which they are constructed (Moore, 1998, pp. 27–30 & 253–254; Moore, 2005, pp. 80 & 93). As with the other features outlined above, a primary selective advantage can be recognised (in the case of septa: 'save punctured hyphae from leaking to death'), but then other functions become possible. Among the first would be to use septa to separate hyphal compartments that differentiate differently. For example, separating sporulating branches from the vegetative hypha, which is quite commonly the case in lower (zygomycete) fungi and is developed to a high level of sophistication in Basidiomycota where hyphal cells can differentiate to contrasting extremes although separated by apparently open dolipore septa (see discussion and illustrations in

Moore *et al.*, 2011, p. 294). Another fungal feature relevant to discussion of the ancient origins of eukaryotes is that:

> Fungal mitotic divisions are intranuclear: in this 'closed mitosis' the division spindle forms inside the nucleus. This is quite different from the 'open mitosis' seen in most animals and plants where the nuclear envelope disassembles and microtubules invade the nuclear space to form the division spindle. When the division spindle is formed within an intact nuclear membrane, progress of the division is more difficult to see and study, but it does not appear to affect the biological consequences of the mitotic division. (Moore *et al.*, 2011, p. 115)

Meiotic divisions in fungi also feature a persistent nuclear envelope. The significance of the persistence of the nuclear membrane as a general characteristic of nuclear divisions in modern fungi is that this is probably a primitive feature and its presence implies that present-day fungi have a more primitive nuclear apparatus than either modern plants or animals, which might further imply that the fungal body plan is also the more primitive.

In animals and plants cell division (cytokinesis) usually occurs in conjunction with nuclear division (mitosis). In fungi, though, mitosis may occur independently of the branching and septation that are the equivalent of cytokinesis in filamentous fungi. The nuclear division spindle is an organelle in its own right. It consists of two organising centres that manage the spindle fibres, which are microtubules, connecting the poles of the division spindle with specialised regions of the chromosomes called kinetochores (protein structures assembled on the centromeres of the chromatids). In animals the two polar organising centres each contain a centrosome made up of two centrioles, and each of the latter is made up of a ring of nine groups of microtubules. In fungi the organising centres are spindle pole bodies (SPBs) which are functionally equivalent to the centrosome but do not contain centrioles. Plants make do without a discrete organising centre. Thus, although the nuclear division mitosis is characteristic of eukaryotes, the three crown groups of eukaryotes involve different organelles in the process.

Another thread of evolutionary logic (in fact co-evolutionary logic) that I believe to be relevant to all stages of the evolution of life is the selective advantage of mutualism (symbiosis) that the fungi, in particular, have exploited (lichens, mycorrhizas, endophytes, anaerobic chytrids in ruminants, and insects that cultivate fungi in their nest gardens), and which can also be traced to more ancient times as the endosymbiotic origin of eukaryotes (Margulis, 2004). As I will argue below, there is a good case for this evolutionary logic being used in even more ancient times during the emergence of living cells; indeed, 'co-operation works' could well be thought of as the First Law of Biology. I will deal with this in Chapter 13. Here, I want to concentrate on the fungal contribution to the emergence of life on Earth, so I will turn back into deep time to make some suggestions about how new experiments and concepts that have become available in the first decade of this century might impact on this. In this past decade or so a great many publications have appeared on topics ranging from interstellar dust particles, via hydrothermal vents in the ocean deeps, to the deep phylogeny of life.

The next chapter starts my review of this and I will begin by briefly painting-in the background by examining how the habitat that has cradled the life we know about was formed, and how it might have acquired the chemical precursors of living systems. There are no fungi (yet) in this part of the story, but there may well be elements of the fungal lifestyle.

THREE

FIRST, MAKE A HABITAT

Around about 5 billion years ago something truly remarkable happened here. I mean here at this unremarkable position in this unremarkable spiral arm of this unremarkable galaxy. At the time this place was a region of interstellar gas and dust, a very tenuous cloud with no particular place to go and no particular thing to do. The remarkable happening was that something disturbed this aimless chaos. It might have been something as simple as a star sweeping past on its own way in its own orbit; it might have been something as dramatic as a star exploding in the vicinity. Either way the gravitational disturbance was enough to give this region of that gas cloud a slight swirl, just the merest touch of concerted angular momentum. And that was enough to start the entire story of life on Earth.

Of course another few hundred million years or so was needed to establish a habitable planet. That initial disturbance set the gas and dust swirling and the resulting interactions caused the dispersed particles to begin to come together into a rotating disc, which retains most of the disturbed cloud's angular momentum. This is a solar nebula: the beginnings of a star and its planetary system. A nebula is an interstellar cloud of dust, gases (and as we will see later, in Chapter 4, organic molecules). Originally, the word nebula was applied to any astronomical object that looked diffuse and cloud-like and many distant galaxies were called nebulae for this reason; indeed, the Andromeda Nebula (= Andromeda Galaxy) was so named even before the nature of galaxies was established. This is old-fashioned usage of the term, which should now be reserved for the clouds of dust and gases that are often star-forming regions.

At the beginning, our own solar nebula was a dynamic and turbulent environment with the simultaneous presence of gas and mineral (silica-rich) dust particles of increasing size range as they aggregate, and energetic processes that include shockwaves, interstellar lightning, and radiation. Within the nebula interstellar gas and dust condense under the influence of the gas dynamics of the cloud and their own gravitational attractions into a presolar mass (the primary for the developing planetary system made up of at least 95% of the total mass) while the remnants form the accretion disc from which interplanetary dust particles, asteroids, comets, planetesimals and planets that orbit the primary ultimately condense. Eventually, the central mass reaches sufficient density to ignite hydrogen fusion and the new star lights up the central part of this solar nebula. The small minority of the total mass remains spread out in the remaining accretion disc. Objects grow in the disc due to the coalescence of dust and ice grains, which might be either survivors from the interstellar medium or newly condensed from materials within the disc itself. It takes about 10 million years to make the first planets.

During this period dust and ice grains first combine through collisions into small clusters and these then gather together, eventually into planetesimals. These are agglomerations which are large enough for gravitational attraction (rather than gas dynamics) to dominate interactions between the planetesimals, and this eventually leads to formation of objects about the size of the present-day Moon. Most of the objects in the disc are eventually swept into the growing star by gas dynamic interactions but some become large enough for gravitational forces (particularly disturbances from Jupiter) to dominate their motions; these are the potential planetary survivors in the system (Lunine, 2006). Depending on the position within the accretion disc of the origin of these embryo planets, they will contain different mixes of prebiotic molecules and volatile species as the initial dust grains also act as condensation sites for the capture of important prebiotic compounds such as water and methane. The outer planets apparently grew fast enough to capture and retain substantial amounts of hydrogen and helium from the solar nebula. The Solar System's comets are believed to be planetesimals accumulated in the Uranus–Neptune region, but scattered to greater distances by encounters with the outer planets; while the asteroid belt is the result of a failed planet, the accumulation of which was frustrated by the gravitational influence of Jupiter.

It probably takes less than about 100 million years for the final planetary masses to be attained, although further accretion from the remaining debris continues for much longer. In the case of the Solar System, the Earth and other solar planets originated from the planetary accretion disc around the young Sun, something like 4.5 billion years ago. This was during the Hadean Eon (4.6 to 3.8 billion years ago) during the early part of which the accretion disc was a chaos of gas, star fragments, dust, comets and planetesimals that

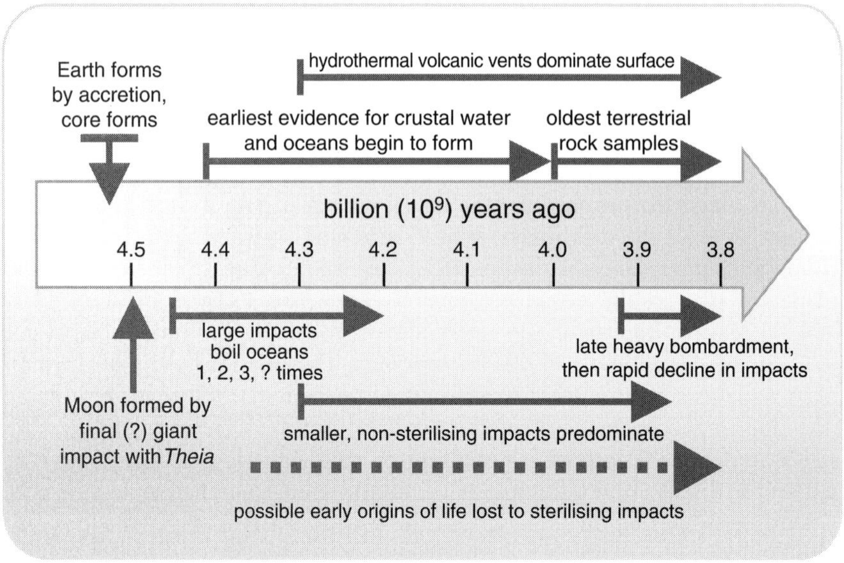

Fig. 3.1 Timeline of the earliest history of the Earth showing the key events in the first 700 million years after formation of the planet. During this period the Earth and Moon were formed and the Earth acquired its water and carbon content. (Based on figure 3 in Lunine, 2006.)

collided and interacted to grow into recognisable planets orbiting the young Sun, and by the end of which the present-day structure of the Solar System had been attained (Fig. 3.1).

Heat generated during accretion induces stratification within the embryo planets; their internal structure separates into:

(a) a planetary core in the centre, made of metals, mainly iron, and may comprise both solid and liquid layers (the metals being maintained in molten form by the heat generated by decay of radioactive isotopes);
(b) a layer surrounding the core, the mantle, which makes up most of the volume of a rocky planet like Earth, and may be segregated into a central layer of flowing rock and an outer layer of plastic but solid rock;
(c) a crust, in a rocky planet, of crystallised and solidified rocky melts which forms the solid planetary surface;
(d) a primitive atmosphere as a layer of gases held in place by the gravitational field of the planetary body. Composition will vary according to the origin of the gases: rocky planets tend to end up with atmospheres composed of gases that have emerged ('outgassed') from their own structure (and on Earth is greatly modified by the planet's biology), gas giants capture most of their atmosphere from the primeval gases in

the solar nebula and most of their mass is made up of such gases (though only the outer layers are counted as 'atmosphere' as the gravitational compaction changes the physical nature of the gases at deeper levels).

If the newly formed planet is a suitable distance from its star to support liquid water at the surface, it is in the so-called 'habitable zone'. Planets with these characteristics are candidates for biological evolution.

Of the chemical elements that play a role in biology today, only hydrogen is primordial. All of the carbon, oxygen, nitrogen, sulphur, phosphorus, metals and other trace elements were produced by nucleosynthesis in the cores of evolved stars or in stellar explosions. Generations of long-dead stars provided the wide range of elements that are found in this Solar System:

> ... most of the chemical elements are synthesized in stars. Helium is made by hydrogen burning in the core during the main sequence and in a shell above the core in the red giant phase. The element carbon is created by helium-burning (the triple-α process), first through core burning and later through shell burning above an electron-degenerate carbon-oxygen core. For massive (more than ten solar masses) stars, direct nuclear burning continues with the production of oxygen, neon, magnesium, silicon and so on, culminating in the synthesis of iron, the heaviest element possible through direct nuclear burning. The other heavy elements, from yttrium and zirconium to uranium and beyond, are produced by neutron capture followed by β decay. (Kwok, 2004)

Then from both interstellar and local (interplanetary) space vast quantities of organic compounds and water are captured to affect the chemistry of the planets throughout their lifetimes. From direct observations and measurements of our own Milky Way Galaxy and other neighbouring galaxies, carbon monoxide characterises the gas in dense interstellar clouds; this gas also contains a wide variety of more complex organic molecules and the interstellar dust grains are also largely composed of organic molecules (see Chapter 4 below).

Uniquely important to the Earth, and to the living things that were to arise on it, was the formation of the Moon. Late in Earth's growth process, about 4.45 billion years ago, a protoplanet roughly the size of present-day Mars (a third or half the size of Earth), which is sometimes called *Theia*, collided with the Earth (Belbruno & Gott, 2005; Hartmann & Davis, 1975; Newsom & Taylor, 1989). Because the Moon is only slightly smaller than planet Mercury, in many respects the Earth–Moon system is a binary system that has several unique features. None of the other terrestrial planets possesses a comparable moon; the tiny moons of Mars, Phobos and Deimos, are probably captured asteroids. The Moon has a high mass relative to the Earth when compared with the satellites of the giant planets, but its bulk density is much lower than that of the Earth or the other inner planets, probably because of its low content of metallic iron. Indeed, samples from the Apollo and Luna missions are strongly depleted in the volatile elements, potassium, lead and bismuth, and relatively enriched in refractory elements such as calcium, aluminium, titanium and uranium, suggesting that the Moon has been fractionated by extreme heat. Also, the Earth–Moon pair has an anomalously high angular momentum compared with the other inner planets, and the inclination of the lunar orbit is unusual. All of these observable features of the present day are consistent with the notion that the Moon originated in a single giant impact: during the final stages of the accretion of the inner planets, when the Earth had attained nearly its present size, a body about the size of Mars, or slightly larger, collided with the Earth and spun out material from which the Moon formed. This was a grazing impact between two bodies that had already differentiated into a metallic core and silicate mantle. The collision disrupted *Theia*, most of its mantle being accelerated away from the Earth and into an orbital disc of rocky debris, and most of its metallic core decelerating, penetrating the mantle and wrapping around the Earth's core. Computer modelling suggests that all of these aspects of the collision could have been completed in about four hours. The debris in orbit probably aggregated into our Moon, perhaps by forming a pair of companion moons first (Jutzi & Asphaug, 2011), which then coalesced to form a partly molten Moon.

The repercussions of this impact for the Earth were very considerable and included complete melting of the Earth's mantle, with about 10% of its mass now being contributed by *Theia*'s mantle. The overwhelmingly important consequence of the collision for the future evolution of life on Earth is undoubtedly that most of *Theia*'s core ends up in the Earth. The particular importance of this event to origin-of-life discussions is that the collision generated the Earth's liquid iron core, and it is this mobile core that protects developing (and extant) life on the planet's surface. The still-spinning molten core of iron generates a magnetic field strong enough to protect Earth from the solar wind, which is a stream of charged particles ejected from the upper atmosphere of the Sun consisting mostly of high-energy electrons and protons. Earth is protected from the solar wind as its magnetic field deflects the charged particles into collisions with atoms in the high-altitude atmosphere and channels electromagnetic energy into the Earth's upper atmosphere, generating the *aurora borealis* and *aurora australis* around the poles because ionised atoms emit photons as they return to ground state.

When first formed, the Moon orbited only about 65 000 km from the Earth (Goldreich, 1966), compared with its present orbital distance of 384 400 km. Within about 5 million years the gravitational gradient caused one side of the Moon to face the Earth permanently; this is called tidal locking (or captured rotation). A tidally locked body takes just as long to rotate around its own axis as it does to revolve around its partner. The Moon's rotation and orbital periods are both just under four weeks (in the present day; they were much shorter originally), so gravitational tidal locking is important biologically because by stabilising the satellite's orbit the lunar month is established and regulates cyclical events such as ocean tides. Tidal locking also influenced the Earth; in particular, the Moon has caused the Earth's rotation to slow gradually over geological time; the Earth day was only 7 hours long when the Moon was first formed (Zahnle *et al.*, 2007). The slower rotation rate reduces temperature variations on the Earth's surface to biologically favourable limits. But tidal locking with the Moon also stabilises Earth's axial tilt. The Earth's axis is tipped over about 23.5° from the vertical and it's our annual orbital motion around that which causes our seasons and provides challenging environments to drive evolution. Similarly, the Moon generates tidal effects in both rocks and water, and the latter also produce variable shoreline environments that spur chemical and biological evolution.

The orbital position of the Earth is crucial to maintaining liquid water on the surface of the planet. The seasonal changes we experience in the present day show how exact that positioning must be. Our seasons result from the planet being angled slightly towards the Sun in summer, or away from the Sun in winter. Just that small rotational shift towards or away from the Sun is sufficient to generate an enormous temperature differential. The highest recorded temperature in Europe (in Seville, Spain) is 50 °C; compared with the lowest recorded European temperature of −55 °C (at Ust-Shchugor, Russia). Had Earth's average orbital distance been very slightly closer to the Sun, say by about one Earth-radius, surface temperatures would be intolerably high all the time. Alternatively, if the planet was a similar distance further from the Sun, it would be in permanent deep freeze. I must confess that this is a considerable simplification. An essential caution is that surface temperatures are mainly influenced by greenhouse gases in the atmosphere (pp. 15–20 in Cockell *et al.*, 2008). Furthermore, the Earth's orbit is elliptical, though the Sun is not at the geometric centre of that ellipse, but at one of its foci; so that Earth comes closest to the Sun during the southern summer and is furthest from the Sun during the northern summer (detailed explanation in Cockell *et al.*, 2008, pp. 6–10). Nevertheless, all these things considered, it remains the case that the planet would be hotter if it were closer to the Sun and cooler if it were further away; and the summer/winter temperature differentials in the two hemispheres imply that the orbital distance between intolerable heat and intolerable cold is rather small.

The giant impact that formed the Moon was just one of several that occurred at about that time (about 4.4 to 4.5 billion years ago) in the Solar System, and only one of many impacts that the Earth–Moon system has suffered in the time since then.

> Craters and ringed basins over 1,000 km in diameter on the Moon, Mercury, Mars and the satellites of the outer planets attest to an early (> 3.8 Gyr [billion years ago]) intense bombardment by a large range of objects. The axes of nearly all the planets are significantly tilted relative to the plane of the ecliptic; the most dramatic example is Uranus, which is lying on its side, probably as a result of a collision with an Earth-sized object. The slow backward rotation of Venus, unique in the Solar System, is most rationally attributed to a late collision with a massive, perhaps Mars-sized object; with a different mass, angle and velocity, the impact might have provided Venus with its own moon. (Newsom & Taylor, 1989, p. 29)

Towards the end of the Hadean Eon the Moon (now 282 000 km from Earth) and Earth suffered a cataclysmic meteorite bombardment (the 'late heavy bombardment'); this was when comets and meteorites rained down on the primordial Earth (Fig. 3.1). A meteorite is any object from space that survives a dive through the Earth's atmosphere and reaches the surface; although rocky, meteorites contain about 5% water and may have brought the bulk of the water on the planet today. Day after day for millions of years the bombardment continued, filling the oceans and scarring the surfaces of Earth and Moon; scars which are still evident on the Moon. Recent evidence from the Mars Reconnaissance Orbiter and the Mars Global Surveyor suggests the late heavy bombardment was caused by debris from an impact between an object 2000 km in diameter and the northern hemisphere of Mars, approximately 3.9 billion years ago (Andrews-Hanna, Zuber & Banerdt, 2008; Nimmo *et al.*, 2008).

Because of the active remodelling of the surface of the Earth the immediate effects of these large impacts are best judged from observation of the Moon or Mars. A striking feature on Mars is the Hellas impact basin in the Southern Hemisphere, which is the deepest impact crater on Mars and one of the largest in the Solar System, and was caused by an asteroid about 1450 km in diameter. This asteroid crater is 9 km deep and 2100 km across. The basin is surrounded by a ring of mountains, made of material ejected by the impact, that rise about 2 km above the surroundings and extend 4000 km from the basin centre. In the Hadean period the Earth would have suffered tens or even hundreds of impacts of this sort, all of them big enough to heat the atmosphere to 1500–3000 K. The shock of a Hellas-scale impact on Earth would vaporise rock and heat the resultant rock vapour and steam atmosphere to 3000 K, though this would cool rapidly to about 1500 K (and the rock vapour would condense at the same time) in a few days and then slowly cool to 500 K in about one year (Fig. 3.2). The surface temperature of the oceans would be raised to approaching 500 K by the impact and would remain at this level for at least 100 years. Similarly, the impact would boil off enough of the ocean to produce a 10 bar atmosphere pressure of steam,

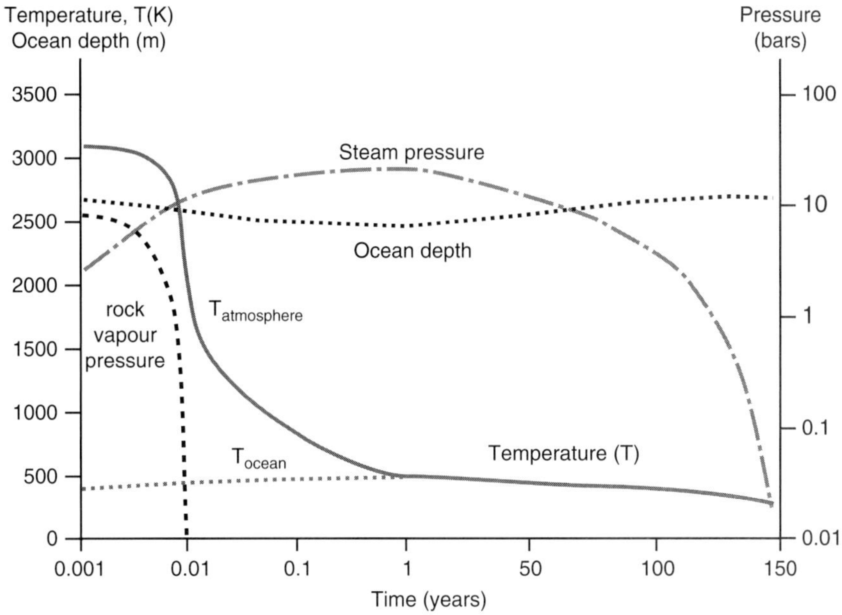

Fig. 3.2 Projected aftermath of the impact of a 1450 km diameter asteroid on Earth. Model is based on impacts such as those that formed the South Pole–Aitken basin on the Moon (2500 km wide, 13 km deep) or the Hellas Planitia on Mars (2100 km wide, 9 km deep); the energy released in such impacts amounts to about 10^{27} joules. The graphs show ocean depth, ocean surface temperature, atmosphere temperature, and the gas pressures of both rock vapour and steam while they are in the atmosphere. Note that the impact heats the entire atmosphere to over 3000 K, which cools on a 0.01 to 0.1 year timescale. By definition, the current oxygen–nitrogen atmosphere of Earth exerts a pressure of 1 bar at sea level. Although not an SI unit, this is a widely accepted unit of pressure, and is often expressed on meteorological charts in millibar (1000 millibar = 1 bar). The equivalent SI unit is the pascal, 1 bar = 100 000 Pa (or 100 kPa). A pressure of one atmosphere (symbol = atm), is now defined as 1 atm = 1.01325 bar. The bar was introduced as a unit in 1909 and is derived from the Greek word *baros*, meaning weight. (Adapted from Nisbet *et al.*, 2007.)

which would be maintained until raining out of the atmosphere about 100–150 years after the impact (Nisbet *et al.*, 2007).

The relevance of this modelling to the origin of life on Earth is that such impacts occurred during a period of Earth's history when the surface was quite capable of giving rise to at least the early stages of life. As we shall see in Chapter 7, it is currently thought that primitive life emerged quite quickly, probably within 1 and 10 million years. So, in the period between formation of the Moon (4.45 billion years ago) and the end of the late heavy bombardment phase (3.8 billion years ago) there was more than enough time for life to emerge, perhaps several times over. However, the heavy impacts described above (and illustrated in Fig. 3.2) are sterilising impacts. Any primeval living thing, and all potentially biogenic organic compounds, on the planet at the time of the impact would have been destroyed. Indeed, the entire atmosphere would have been chemically reset to something similar to that which first formed after the formation of the Moon.

Of course, asteroids are not the only objects that could cause planet-changing impacts. Impacts of comets on Earth would have been dramatic, too. It has been calculated that at least 100 comets have collided with the Earth since the formation of the planet (Oró, 1961) and has been suggested that the most important consequences of cometary impacts would have been the accumulation on the planet of large amounts of carbon compounds and large amounts of water, although this idea has been criticised on the grounds that cometary carbon compounds might not be expected to have the thermal stability to survive impact. However, recent experimental studies have demonstrated that amino acids and other carbon compounds found in comets can indeed survive ballistic impacts (the pressure at the impact suppressing the adverse effects of heat) and, further, show that the impact itself provides energy to synthesise new organic compounds (Blank *et al.*, 2001). Such experiments support the idea that comets might have been a source of organic matter for the early Earth that may have influenced or promoted the origin of life. Comets, asteroids and meteorites that cause major impacts, together with a steady 'rain' of interplanetary dust particles that reach the surface of Earth without experiencing catastrophic impact, are all external sources of potentially biogenic organic compounds that will be discussed and quantified in Chapter 4.

There is no geological record for the Hadean Eon, because the surface of the Earth has been remodelled so completely in the past 3.9 billion years, but it is possible to give an impression of the conditions that existed then. In his foreword to Hazen (2005), David Deamer describes the environment of the surface of this early Earth as follows:

> Imagine that we could somehow travel back in time to the prebiotic Earth, some 4 billion years ago. It is very hot – hotter than the hottest desert today. Asteroid-sized objects bombard the surface. Comets crash through the atmosphere – no oxygen yet, just a mixture of carbon dioxide and nitrogen – and add more water for the eventual globe-spanning ocean. Landmasses are present, but they are volcanic islands resembling Hawaii or Iceland, rather than continents. Imagine that we are standing on one such island, on a beach composed of black lava rocks, with tide pools containing clear seawater. We can scrutinize that water with a microscope, but there is nothing living to be seen in it, only a dilute solution of organic compounds and salts. If we could examine the mineral surfaces of the lava rocks, we would see that some of the

> organic compounds have formed a film adhering to the surface, while others have assembled into aggregates that disperse into the seawater. (Hazen, 2005, p. x)

To this I would add a few more features. At this time the Earth day was about 14.5 hours long and the Sun was only a little more than 70% as bright as today. Earth's original hydrogen and helium atmosphere had escaped the planet's gravity early in the Hadean, but outgassing from volcanoes and incoming comets and meteorites created an early atmosphere of water vapour, methane, ammonia and, especially, carbon dioxide. As a consequence, formation of carbonate minerals started at this time. Note that the absence of oxygen in the atmosphere means that there was no filtering of ultraviolet radiation reaching the surface. In today's atmosphere we benefit from the protective effect of the ozone layer (itself protected from being stripped away by the solar wind by our magnetosphere). The ozone layer is crucially important to biology; UV at wavelengths shorter than 280 nm (also known as 'germicidal UV' or UV-C) is entirely screened out by ozone at around 35 km altitude. Radiation with a wavelength of 290 nm (UV-B, which is most damaging to DNA), has an intensity at Earth's surface only 10^{-8} of that at the top of the atmosphere, and this is thanks mainly to high-altitude ozone.

When the late heavy bombardment calmed, about 3.8 billion years ago, the surface of the Earth stabilised, changing from mostly molten rock to mostly solid rock, a surface on which water could condense to liquid form, and million-year-long rainstorms could fill the oceans (Figs. 3.1 and 3.2). This is the natural environment, the essential habitat, for the origin of life; for more details and explanation about it see Cockell *et al.* (2008, particularly their chapter 1). Our next task is to assess how the chemical precursors of life, its building blocks, came to this habitat.

FOUR

THE BUILDING BLOCKS OF LIFE

Our present understanding is that the Universe is between 12 and 15 billion years old and recent experiments and observations suggest that for almost all of that time most of the elements that we now know in the Periodic Table have been present and there has also been an abundance of spontaneously synthesised molecules, most of these being organic molecules. These exist in the interstellar medium of our own Milky Way Galaxy and other galaxies, and in our Solar System.

Max Bernstein (2006) starts the abstract of his article on prebiotic materials with these sentences:

> One of the greatest puzzles of all time is how did life arise? It has been universally presumed that life arose in a soup rich in carbon compounds, but from where did these organic molecules come? (Bernstein, 2006)

Before showing how Max Bernstein answered his own questions, I want to ask (and answer) the question where did these 'universal presumptions' come from? As with many aspects of modern biology, we can look back with expectation of enlightenment to the writings of Charles Darwin.

> Although Darwin's *Origin of Species* is still widely believed by many to refer to the origin of life, this was not a question the book addressed. His book was instead about where existing species come from. The short answer is that they are genealogically descended in an unbroken reproductive series from earlier species. But what did Darwin think about the origin of life? In the *Origin of Species* he wrote 'I should infer from

> analogy that probably all the organic beings which have ever lived on this earth have descended from some one primordial form, into which life was first breathed.' Although a few years later in 1863 Darwin wrote to his friend the botanist Joseph Dalton Hooker: 'I have long regretted that I truckled to public opinion and used Pentateuchal term of creation, by which I really meant "appeared" by some wholly unknown process.—It is mere rubbish thinking, at present, of origin of life; one might as well think of origin of matter.' Yet despite these cautious protestations we can clearly glean from his occasional references to the origin(s) of life that Darwin believed that life arose by purely natural causes as simple micro-organisms in an aquatic environment on Earth. (van Wyhe, 2010)

Undoubtedly the most specific description appears in a letter which is often quoted in origin-of-life discussions. His son Francis records that his father wrote in 1871 [in a letter to J. D. Hooker]:

> It is often said that all the conditions for the first production of a living organism are now present, which could ever have been present. But if (and oh! what a big if!) we could conceive in some warm little pond, with all sorts of ammonia and phosphoric salts, light, heat, electricity, &c., present, that a protein compound was chemically formed ready to undergo still more complex changes, at the present day such matter would be instantly devoured, or absorbed, which would not have been the case before living creatures were formed. (Darwin, 1887)

Fenchel (2002) points out that the kernel of the idea that derives from Darwin's notion is that:

> ... one goes from a random mix of essential but disordered components to the first self-replicating molecule, and from there to the last common ancestor is a simple matter of natural selection ... In other words, scientific origins of life scenarios tend to put the pieces of life in proximity to one another, add some kind of energy and figure that eventually something or some process that can reproduce itself will arise. (Fenchel, 2002, see his chapter 3)

The visualised location may be a warm little pond as Darwin imagined, organic scum on a rocky shore (Follmann & Brownson, 2009), or a deep, hot oceanic volcanic vent as envisaged by Wächtershäuser (2006), but the primordial process is a chemical evolution taking place in a chemical broth that already contains many of the essential components we would recognise as contributing to a recipe for the generalised living organism alive today (meaning amino acids, carboxylic acids, purines, pyrimidines, sugars, as well as a range of common inorganic salts, and trace elements too). The designation of this as a primordial 'soup' originates from J. B. S. Haldane's description of the prebiotic ocean as 'a hot thin soup' (Haldane, 1929).

Now we need to address Max Bernstein's question about the origins of the ingredients of the soup and it seems only fair to quote Bernstein's own definitions because they are really helpful:

> Defining 'organic' and 'reduced': when I refer to organic compounds, I mean those composed primarily of carbon, but may also contain nitrogen, oxygen and other elements. These are the kinds of molecules from which we and all living things are made, as opposed to carbon in the form of carbonate rocks, which is considered 'inorganic'. Technically 'reduced' carbon is that bearing hydrogen atoms, such as in methane (CH_4). Oxidized carbon is that such as in carbon dioxide (CO_2) or carbonate rock, where all of the bonding is satisfied by oxygen atoms. For the purposes of being good prebiotic molecules partially oxidized carbon species (e.g. ketones, $>C=O$) will be of use in making bigger, more complicated, and biologically important compounds. A 'reduced' gas is rich in hydrogen, or compounds that bear hydrogen, such as methane (CH_4) and ammonia (NH_3). Reduced gas mixtures produce more complex organic molecules than do oxidizing ones which is why the oxidation state of the early Earth's atmosphere is so important. (Bernstein, 2006)

Probably the most surprising thing about organic compounds is that the Universe seems to be full of them; Ehrenfreund & Cami (2010) put it this way:

> Astronomical observations have shown that carbonaceous compounds in the gas and solid state, refractory and icy are ubiquitous in our and distant galaxies. Interstellar molecular clouds and circumstellar envelopes are factories of complex molecular synthesis. A surprisingly large number of molecules that are used in contemporary biochemistry on Earth are found in the interstellar medium, planetary atmospheres and surfaces, comets, asteroids and meteorites, and interplanetary dust particles. (Ehrenfreund & Cami, 2010)

Hydrogen is the most abundant element in the Universe, of course, and setting the elemental abundance of hydrogen to unity, the relative elemental abundances of the other major component elements of organic compounds (those we can call primary biogenic elements) are: carbon, 4×10^{-4}; oxygen, 8.3×10^{-4}; nitrogen, 1×10^{-4}; sulphur, 1.7×10^{-5}; phosphorus, 3×10^{-7} (Ehrenfreund, Charnley & Botta, 2005). These numbers mean that for every carbon atom in the Universe there are 40 000 hydrogen atoms; for every oxygen atom there are 83 000 hydrogens; and for every nitrogen atom there are 10 000 hydrogens, etc., and they also mean that carbon comprises 0.04% of the mass of the Universe (etc.). I will outline the potential sources of prebiotic organic molecules that have been proposed to be input to Earth from interplanetary and interstellar space in this chapter and those that could be synthesised on the planet in Chapter 6. In later chapters I will indicate how those chemicals might depend on the physical state of the planet, the chemistry of the atmosphere, and the phylogenetic sequence that can be presumed as a result of these considerations.

Molecules do occur in the stars themselves but these tend to be simple diatomic molecules, and are found particularly in sunspots. So we start far from Earth in interstellar space where organic molecules occur in the interstellar medium from which new stars form and to which evolved stars return the products of nucleosynthesis when they explode. Only atoms (neutral and ionised) are found in the majority of interstellar space. Molecules are concentrated in 'clouds', some of which are partially transparent to optical and ultraviolet light and to radio waves which can be detected on Earth, allowing many familiar organic molecules to be identified by spectroscopy at a broad range of wavelengths. Wavelengths near to the UV end of the spectrum are used to identify the three free radicals, CH•, CHC• and CN• (in this notation the dot '•' signifies that the chemical is the free radical of the indicated molecule, free radicals have unpaired electrons and are highly chemically reactive). The visual or infrared wavelengths identify non-polar molecules such as C_2, methane, acetylene and polycyclic aromatic hydrocarbons (or PAHs), but radio wavelengths have mainly been used and these can identify ammonia and water and many organic molecules with great accuracy. Radio astronomy can assign identifications with near certainty because interstellar molecular gas is typically cold and radio lines are often extremely sharp; in favourable cases it is possible to match astronomical lines to laboratory standards within a few parts in 10^7. In the interstellar medium of the Milky Way Galaxy more than 100 organic molecular species have been observed and identified. These observations are crucially important because the solar nebula from which the primitive Solar System emerged was formed by collapse of such an interstellar cloud, so from the very start the material that made the Solar System contained these molecules and they must have contributed to the objects such as the early Earth, asteroids and comets that were formed by accretion of the dust particles from the solar nebula.

Thaddeus (2006) lists 135 molecules that had been identified by radio astronomy in the interstellar gas and circumstellar shells (the spherical shell of space surrounding a star, rich in newly formed elements, dust grains and molecules). This list includes (in addition to compounds already mentioned) acetone, benzene, ethanol and ethylene glycol. The inventory has grown at the rate of an average of five molecules identified per year for the past 40 years and the largest molecule in the 2006 list is a 13-atom cyanopolyyne carbon chain [cyanodecapentayne, $H(C{\equiv}C)_5CN$] with a molecular weight of 147. More recently, ethyl formate (C_2H_5OCHO) and *n*-propyl cyanide (C_3H_7CN) have been added to the list (Belloche *et al.*, 2009).

In addition, there is a great deal of evidence for significantly larger organic molecules being important components of the interstellar gas, especially

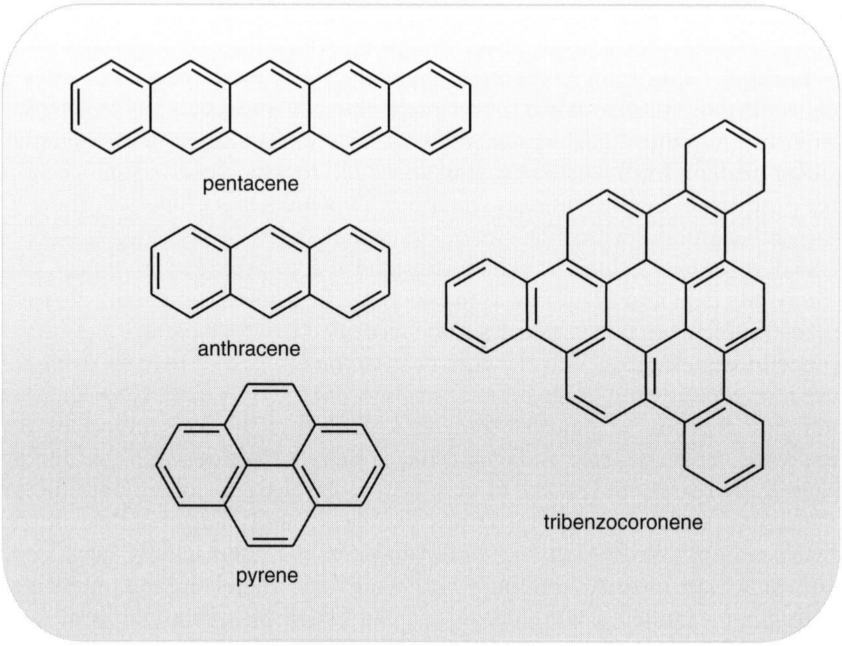

Fig. 4.1 Examples of polycyclic aromatic hydrocarbons (PAHs): pentacene, anthracene, pyrene and tribenzocoronene. PAHs are probably the most abundant organic gas phase molecules in space. They are observed ubiquitously in our galaxy and beyond through their spectroscopic signature in the near- and mid-infrared bands. They seem to be stable and abundant components in diffuse clouds and other space environments. (From Ehrenfreund, Charnley & Botta, 2005.)

polycyclic aromatic hydrocarbons or PAHs, although precise identifications of individual chemical species are few (Fig. 4.1). PAHs generally have weak or non-existent radio spectra and are identified by infrared and UV emission bands. Anthracene ($C_{14}H_{10}$) and pyrene ($C_{16}H_{10}$) were the first to be identified (in 2004; Vijh, Witt & Gordon, 2005) in a (protoplanetary) star nebula called the Red Rectangle nebula [visit this URL: www.esa.int/esaCP/SEMHBNGHZTD_Feature Week_0.html].

It is thought that in the turmoil of currents in the stellar winds of nebulae, in the remnants of carbon-rich stars, PAHs could grow into molecular particles of a million or more atoms from reactions between smaller unsaturated hydrocarbon molecules and radicals. PAHs would be stable since they efficiently re-emit at infrared wavelengths the energy they absorb from the interstellar ultraviolet radiation flux. Because of their abundance PAHs are potentially important prebiotic molecules. They are made up of 6-membered rings of carbon atoms, linked in various combinations, with hydrogen atoms around their edges. If they came to Earth, in the regular rain of interplanetary dust into the atmosphere or in larger objects like meteorites or comets, the hydrogen atoms could be stripped off in water and replaced with hydroxide (-OH) under the effect of solar UV radiation. This increases their solubility and enables them to self-organise; because the interior (hydrophobic) parts of the molecule are relatively flat, they can stack one on top of the other. In this configuration the hydrophobic interaction stabilises the stack which becomes a potential amphiphilic macromolecular chemical reactor that might aid biologically important reactions such as the binding and assembly of DNA bases (discussed at greater length by Hazen, 2005, chapter 17, pp. 224–225).

Perhaps the most remarkable compounds so far found in nebulae are C60 and C70 fullerenes (or 'buckyballs'; see www.bristol.ac.uk/Depts/Chemistry/MOTM/ buckyball/c60a.htm); the two molecules amounting to a few per cent of the cosmic carbon in the region (Cami *et al.*, 2010; Ehrenfreund & Foing, 2010). PAHs and fullerenes, both of which are large carbon molecules that contain many aromatic rings, may be formed by the photochemical processing of hydrogenated amorphous carbon (García-Hernández *et al.*, 2010).

Familiar carbon molecules are very widely distributed. Strong signals for carbon monoxide (CO) and hydrogen cyanide (HCN), both intermediates for the prebiotic synthesis of biogenic molecules, are observed along the entire plane of our galaxy. Indeed, CO is widespread in other spiral galaxies, where the molecular clouds follow the spiral arm aggregations of visible stars. The galactic clouds of simple gases are accompanied by dust grains.

> The difference in mass between the largest molecules and the smallest interstellar dust grains is now so small – only about one-half an order of magnitude – that it is plausible to postulate a continuous transition from molecules to grains, and to suppose that these two constitute a single continuous population of chemically bonded structures ... the largest grains for which observational evidence exists: objects about 0.4 mm in size with the order of 10^{10} atoms ... The availability of free energy in the interstellar gas allows, in principle, the assembly of structures of arbitrary complexity; so, it is plausible that many of the interstellar grains have very specific structures, and are in fact large molecules. (Thaddeus, 2006)

In these interstellar gas clouds constructive free energy is available in the form of heat (only 100–500 °C is required), ultraviolet light and electric discharges; although high temperatures may not be so important because: 'Below 0 °C, hydrogen cyanide (HCN) and water exist in an HCN-rich eutectic state (that is, freezes at a lower temperature) which appears ideally suited for condensation reactions and freezing-out of heterocyclic products' (Follmann & Brownson, 2009). Consequently, all that may be necessary to synthesise biogenic molecules can be found in the gas and dust clouds that make the nebulae within which stars and planets are formed – throughout this galaxy and every other.

Interstellar matter provides the raw material for the formation of stars and planets and this includes a great many carbon compounds which are recycled during solar system formation. When the planetary system emerges from the turbulent environment of the solar nebula these materials of interstellar origin are confined to the developing Solar System, contained within the remnants of the accretion disc as interplanetary dust particles, asteroids, comets, planetesimals and planets that orbit the central star.

Asteroids currently located in the outer regions of the Solar System are probably the parent bodies of the many carbonaceous chondrites on Earth, which are a common type of meteorite that contains chondrules, the most famous example being the Murchison meteorite, which is described below. Chondrules are roughly spherical inclusions, 0.5 to 2 mm in diameter, which are usually composed of iron, aluminium, or magnesium silicates. They are among the oldest objects in the present-day Solar System with an age of about 4.6 billion years, having been formed when interstellar dust in the solar nebula was heated to temperatures high enough to become molten droplets, which then resolidified and aggregated into asteroids. Carbonaceous chondrites found on Earth may contain as much as 5% organic material, which is composed of a large variety of extraterrestrial amino acids, sugars, purines, pyrimidines, and other species of potential significance to the most primeval biology (Martins, 2011). These compounds provide an illustration of the range of chemical reactions and conditions in the early Solar System. Different mechanisms of formation and three different sources (interstellar, nebular or parent body) contributed to the inventory of organic molecules observed in meteorites collected today.

Some of the organic content of carbonaceous chondrites may be of similar age to the chondrules, although it is likely that a considerable proportion was formed within the parent asteroid in the Solar System. Many extraterrestrial materials in meteorites show evidence of having experienced liquid water in their parent body before they landed on Earth (Bernstein, 2006). Particularly interesting combinations of analytical observations and chemical experiments have recently suggested that the relative enrichment in the amino acid L-isovaline in comparison with its isomer D-isovaline (see legend of Fig. 2.2 for explanation), which is observed in the Murchison and other meteorites, resulted from alterations in the molecules that could only have happened in liquid water. This implies that the asteroids from which these meteorites fragmented (the parent bodies) were large enough to be sufficiently heated by radionuclide decay for liquid water to exist for long periods in internal crevices. If the temperature was high enough to maintain the aqueous phase for extended periods, the isomerisation could take place, protected from the adverse effects of solar radiation by being shielded within the rock. It is significant that present-day biology shows an extreme (almost absolute) bias towards the L-isomer of amino acids and these meteorites also show the same isomer asymmetry (that is, the L-isomer predominates in extracts of the meteorite). The authors suggest that amino acids delivered to Earth by asteroids, comets, and their fragments could have biased the Earth's prebiotic portfolio of organic compounds with L-isomer amino acids and D-isomer sugars before the origin of life, which is now exhibited by present-day biology (see discussion in Glavin & Dworkin, 2009; Ehrenfreund & Cami, 2010).

Carbonaceous meteorites also contain a wide range of pyrimidine and purine bases on which all present-day organisms depend for their nucleic acids: that is, their RNA and DNA (Callahan *et al.*, 2011). Importantly, this study included some experimental chemistry and showed that an identical collection of purine nucleic acid bases and their analogues that were observed in extracts of the meteorites were generated in chemical reactions with ammonium cyanide. The authors state that their results demonstrate that the purines detected in meteorites are consistent with these being the products of an ammonium cyanide-based chemical synthesis that took place within the asteroid parent bodies (Callahan *et al.*, 2011). This strongly supports an extraterrestrial origin for these organic compounds and expands the prebiotic molecular catalogue that was available when the first genetic molecules were constructed.

Observations of the comas of comets (the nebulae that surround the comet head) show that comets are rich in organic compounds. Volatile organics, which can be observed after their sublimation into the coma, include many chemicals also present in the interstellar medium, and it is thought that most of the Earth's volatiles may have been supplied by a late bombardment of comets and carbonaceous meteorites, which had been scattered from the outer Solar System by the gravitational disturbance of the gas giant planets.

Clearly, throughout the aggregation/accretion processes that formed the components of the Solar System, all the organic compounds that were present in the original nebula were included into the objects that resulted. In addition, chemical processing of these organics continued within those objects to form an ever wider range of biogenic molecules in interplanetary space. The organic compounds do not survive all of this 'processing'. The accretion process gradually builds up heat so in their early stages rocky planets, like Earth, were hot enough for the rock to be molten to the surface, and indeed even vaporised to form a hot silicaceous atmosphere. In these circumstances infalling water-ice in asteroids or comets is vaporised, but remains in the atmosphere, and existing organic compounds in or on the infalling objects are chemically altered (pyrolised) by the heat produced by friction with the atmosphere, but the products of that pyrolysis, even if elemental, contribute to the chemical inventory of the planet and its atmosphere.

As the planetary surface cooled, the surface rock solidified and eventually water could condense on the surface (Fig. 3.1); in these circumstances a high proportion of the extraterrestrial materials delivered to the surface of the young Earth could survive their arrival. Comets rich in organic compounds will have effectively delivered carbonaceous solids and volatiles to the surface of the young planet as well as large quantities of water. Many meteorites, particularly the carbonaceous chondrites described above, contain prebiotically significant organic material. Perhaps the most celebrated example is the Murchison meteorite, which fragmented over the town of Murchison, Victoria, in Australia, on 28 September 1969. Over 100 kg of fragments of this meteorite, with individual masses up to 7 kg, have been collected.

By definition, compounds detected in this meteorite and similar ones survived their arrival through our current atmosphere; those compounds include amino acids, sugars, purines, pyrimidines, carboxylic acids, sulphonic and phosphonic acids, and aliphatic and aromatic hydrocarbons. The major carbon component of meteorite samples is a macromolecular organic fraction; and the dominant form of carbonaceous material that will be delivered this way is likely to be aromatic in nature. This extraterrestrial material arrived in large quantities during the early history of our Solar System. It has been estimated (Chyba, Brookshaw & Sagan, 1990; Chyba & Sagan, 1992) that such sources delivered about 10^9 to as high as 10^{13} kg of carbon per year to the Earth during the 600-million-year late heavy bombardment phase that ended 3.9 billion years ago; that's a total of 6×10^{12} to 6×10^{16} metric tonnes (a quick 'Google' of that quantity using the lower estimate reveals that it's about equal to the current estimate of the Earth's coal reserves, or is about ten times the estimate of the total terrestrial biomass in Chyba *et al.*, 1990). The estimate for the present day is that 40 000 tons of extraterrestrial material fall to the Earth's surface every year. Assuming (conservatively) that this amount is representative of the rate of infall since the end of the late heavy bombardment and that 10% of the mass is carbonaceous (the Murchison meteorite is 12% carbon) then a total of 1.6×10^{13} tons of extraterrestrial carbon compounds have arrived on Earth over the past 3.9 billion years; that's a lot of potential biomass.

That this organic material could survive delivery to the Earth sufficiently for its prebiotic biological potential to be realised has been doubted over the years because it was considered more likely that the organics would have been destroyed by the heat liberated during descent through the atmosphere and on impact. Those who believe this maintain that the most likely source of extraterrestrial organic compounds are the interplanetary dust particles that the Earth's atmosphere collects during its orbit and which settle out of the atmosphere with minimum frictional and impact heating. However, the organic compounds in the Murchison meteorite obviously survived the landing experiences of that object and experiments have been carried out to investigate directly the effects of ballistic impacts on amino acids (Blank *et al.*, 2001). All the amino acids tested survived their impacts and some of the amino acids combined to form dimers, possibly via peptide bonds, supporting the notion that comets could have been a source of organic materials to the very early Earth (Blank *et al.*, 2001; and see Chapter 7).

FIVE

AN EXTRATERRESTRIAL ORIGIN OF LIFE?

In Chapter 4, I stated that 'interstellar matter provides the raw material for the formation of stars and planets' and it should now be evident that interstellar matter may well have also been important starting material for the origin of life on Earth (Bernstein, 2006; Chyba *et al.*, 1990; Ehrenfreund & Cami, 2010; Ehrenfreund *et al.*, 2002). In fact, in Hazen's phrase:

> The bottom line is that the prebiotic Earth had an embarrassment of organic riches derived from many likely sources. Carbon-rich molecules emerge from every conceivable environment. Amino acids, sugars, hydrocarbons, bases – all the key molecular species are there. (Hazen, 2005, p. 127)

This being the case, and as there are so many plausible mechanisms for the synthesis of biogenic compounds *in situ* on planet Earth, it is surprising that so much attention has been given to the possibility that life came to Earth ready formed rather than originating here. The panspermia hypothesis claims that life exists throughout the Universe, with microbes drifting through interstellar and interplanetary space, transmitting life to the next habitable body they encounter. According to this view the Earth of long ago was colonised by microbes that had somehow escaped from their home planets to drift across the vast distances between the stars until they arrived on that primeval and sterile Earth. Panspermia should not be confused with pangenesis; the latter was Charles Darwin's conjectural mechanism to explain heredity (he argued that inheritance depended on particles produced by each organ being transmitted from parent to offspring).

The panspermia hypothesis was formulated long before the construction of the first radio telescopes, on which so much of our knowledge of interstellar chemistry depends; in fact it can be traced back to the classical Greek astronomer and mathematician Aristarchos of Samos (310 to about 230 BC) whose main claim to fame is that he was the first person to place the Sun at the centre of the Solar System and the planets in their correct order of distance around the Sun, although, of course, this act of genius was buried under Aristotle's Earth-centred dogma until Nicolaus Copernicus rescued the truth 1800 years later with his comprehensive heliocentric cosmology in 1543. In the early years of the twentieth century, panspermia was revived by the Swedish physicist and chemist Svante Arrhenius who suggested explicitly in 1908 that life had originated somewhere else in the Universe and in fact permeates all of the cosmos (Arrhenius, 1908). Arrhenius had been awarded the Nobel Prize in Chemistry in 1903 for his 'electrolytic theory of dissociation' (www.nobelprize.org/nobel_prizes/chemistry/laureates/1903/#) so his opinions carried considerable weight and, as Fenchel (2002) puts it, the panspermia hypothesis:

> ... haunted most of the 20th century. Supporters include such prominent persons as Francis Crick (who together with James Watson and others discovered the structure of DNA), although he later claimed that it was meant only as a provocation to show how difficult it is to understand the origin of life. The astronomer Fred Hoyle (also known as an author of science fiction novels) seemed, however, to have taken the Panspermia hypothesis seriously." Fenchel (2002, chapter 3, p. 12)

Fred Hoyle certainly did take panspermia seriously and with several collaborators (especially Chandra Wickramasinghe) published a great deal on the topic in the last 20 years of the twentieth century and, following Hoyle's death in 2001, Wickramasinghe continued the work into the twenty-first century (e.g. Wickramasinghe, 2010). It is no coincidence that panspermia grew in stature as a hypothesis (or, in Fenchel's words: as a 'haunting') across the latter half of the twentieth century because this was the golden era of so many aspects of human endeavour that looked outwards into space:

(a) The space race to get manned space vehicles into Earth orbit and then on to Moon landing.
(b) Soviet Russia's and NASA's planetary explorations, particularly NASA's Grand Tours around the Solar System.
(c) Developing and ever improving radio astronomy.
(d) Improvements on optical astronomy from Earth-based telescopes and from those in orbit (the Hubble Space Telescope was placed into orbit in 1990 and its optics were corrected in 1993).

(e) And, perhaps most influential of all, the impact of photographic images from space; from the Hasselblad images of the Apollo Moon missions to the Hubble Ultra-Deep Field images of the Universe of 13 billion years ago.

All of these efforts contributed to new, exciting, science which appeared to be providing evidence for panspermia. For example, Jay Melosh (1988) was able to compute whether large meteorite impacts on Earth may eject rocks into space that could eventually fall on Mars or other planets in the Solar System. He concluded that this could indeed happen and that although transit times could be in the range of millions of years:

> Planets of the Solar System should therefore not be thought of as biologically isolated: from time to time large impacts may inoculate Mars and the other planets of the inner Solar System with a sample of terrestrial life." (Melosh, 1988, p. 688)

Add to such theoretical analyses a growing number of experiments on the ability of microbes to survive on space vehicles (Horneck, 1996, 1999) and it is easy to understand the enthusiasm with which people approached ideas involving microbes in outer space.

Now that we have a diverse and vibrant biology on Earth, I do not deny the possibility that we may share parts of it with other planets in the Solar System; either by bacterial or other spores drifting into space from the highest reaches of the atmosphere, by more explosive ejection as a result of meteorite impacts, or indeed resulting from contamination by inadequately sterilised parts of our own spacecraft. However, I see this as cross-contamination from Earth with life that originated on Earth and I do deny that panspermia has any validity at all as an explanation for the origin of life on Earth. In what I intend to be a brief description, I will here concentrate my attention on the Hoyle/Wickramasinghe story. If you want to get additional information about some of the more imaginative variations on the panspermia theme, including the notion that life on Earth developed from garbage jettisoned by an alien space ship on the sterile primeval planet; or directed panspermia, which is the theory that organisms were deliberately spread to Earth and elsewhere by an advanced extraterrestrial civilization located on another planet; or indeed, that Earth microbes might be distributed by humans to other planets by fleets of specially designed spacecraft, see Crick & Orgel (1973) and view the URL www.panspermia-theory.com/. Detailed reviews of the history of origin of life theories can be found in the following books: Davies (2006, chapters 1, 9 and 10); Dyson (1999, chapters 1 and 2); Fenchel (2002, chapter 3); Luisi (2006, chapters 1 and 2); Lurquin (2003, chapter 6); Oparin (1957a, chapters 1 and 2). The paper by John van Wyhe (2010) is also highly recommended, both for its historical content and for the remarkable quote from Charles Darwin: 'Almighty God! What a wonderful discovery!'.

The Hoyle/Wickramasinghe story started in the 1960s with conventional astronomical research on the nature of cosmic dust, which is composed of particles in space which are a few molecules to 300 μm in size. Cosmic dust plays a part in the early stages of star and planet formation; in our own Solar System sunlight scattered by a cosmic dust cloud is responsible for the zodiacal light seen from Earth, and there are planetary dust rings around Jupiter, Saturn, Uranus and Neptune, and comets and meteorites also contain cosmic dust. At the start of their research the consensus of opinion was that cosmic dust particles were ice grains similar to the ice crystals that exist in the clouds of Earth's atmosphere. Hoyle and Wickramasinghe showed that this was not the case and were the first to show convincingly that interstellar dust was comprised largely of the element carbon. There is no problem with this analysis and I have discussed in Chapter 4 the widening range of organic carbon compounds that are being identified in this dust. The techniques being used for these identifications today were not available to Hoyle and Wickramasinghe. Instead, largely from a comparison of the size range of present-day bacterial cells with that of cosmic dust particles, and the fact that bacteria are carbonised by the sorts of conditions that cosmic particles might endure, they concluded that cosmic dust is comprised of bacteria and bacterial degradation products. They assembled a range of biological and medical observations into arguments they thought were 'already sufficiently compelling to assert that the thesis of life being cosmic was all but proved' (Wickramasinghe, 2010, p. 128), which was published under the title *Proofs that Life is Cosmic* in 1982 (Hoyle & Wickramasinghe, 1982). At the time of writing, this document is still available at the URL: www.panspermia.org/proofslifeiscosmic.pdf but a glance at the section subheadings shown in the following list will provide sufficient impression of the contents:

(a) Atmospheric entry of microbes, showing that microbes can survive entry to Earth.
(b) Bacteria; their amazing properties of radiation resistance and survival under space conditions.
(c) Comets; their role as amplifiers and distributors of life in the galaxy.
(d) Diseases; the still contentious connection between comets and epidemics.
(e) Evolution; showing that the evolution of life on Earth requires an open system, including periodic additions of pristine extraterrestrial genetic material. These are inserted like subroutines into a computer programme to be used whenever an opportunity arises.
(f) Interstellar dust; the properties of which show a connection with bacteria and bacterial degradation products.

(g) Meteorites; arguments about microfossils in meteorites.
(h) Origin of Life; elusive concept.
(i) Planets; evidence that life is widely present in the Solar System. (Hoyle & Wickramasinghe, 1982)

Thus, the essential ideas of this theory are that the sterile primeval Earth was inoculated with live bacteria in cosmic dust grains caught up into comets in the young Solar System. Hoyle & Wickramasinghe (1982) also maintain that genetically active molecules (bacteria, viruses, DNA or RNA) originating from extraterrestrial cosmic dust have been incorporated into the evolution of most organisms through geological time as stimuli to further evolutionary adaptations. Further, they view this last process as continuing into the present day by being manifest as viral and bacterial infectious diseases affecting humans and their animals and plants in historical time. But the best they can say about the origin of life is that it is an elusive concept.

I remain unconvinced. It is, in my view, wrong to decide that because we cannot come up with an explanation of how life originated we will assume it came ready formed from elsewhere. The premise of this approach is completely unscientific; more like magic, and while I am impressed by the magician's skill and dexterity, I know it's just hocus-pocus.

I want it to be clear that there is now enough evidence available to say that panspermia from Earth is a possibility. Experiments have demonstrated that the spores of present-day bacteria are sufficiently resistant to temperature and pressure shocks, and radiation exposure, to survive in space on artificial satellites and other spacecraft for periods of several years, time enough to cover transit between planets in the Solar System. It is also evident that rocks from Mars have been found on Earth. Assuming that the reverse transit is also feasible, now that we have life everywhere on Earth it is possible that it could be exported (or even could have been exported already) from Earth to other planets in the Solar System (see discussion in Lurquin, 2003, chapter 6, pp. 154–158). But it remains the case that there is no evidence of life of any sort anywhere other than on this planet Earth. Nevertheless, Chandra Wickramasinghe remains a staunch defender of panspermia and recently wrote:

> Many of our earlier arguments under category [Evolution] have been developed and amplified ... to the point of becoming decisive ... cycles of prediction-verification- re-affirmation to put a theory or hypothesis to ever more stringent tests. Needless to say it has led to a veritable list of successes and confirmations over the past three decades, implying consistency, nay proof, of the hypothesis of panspermia ... confirming the panspermia hypothesis has been enormously strengthened in recent years. The spectroscopic identification of interstellar dust and molecules

> in space, which was our starting point in the 1970s, has come into much sharper focus. Their biochemical relevance is now widely conceded, although a fashion remains to assert without proof that we are witnessing the operation of prebiotic chemical evolution on a cosmic scale. If biological evolution and replication are regarded as the only reliable facts, life always generates new life, and this must surely be so even on a cosmic scale. Prebiology, whether galactic or planetary, remains an unproven hypothesis that fails the test ... [prediction–verification–re-affirmation] ... It is in the author's view a mistaken remit of modern astrobiology to seek an origin of life everywhere where conditions appear to be congenial. The genetic components of life, no matter where it first arose, are mixed on a Galaxywide scale. Life was most likely to have been first introduced to Earth during the Hadean epoch by impacting comets billions of years ago, thereby establishing our cosmic ancestry. However, the precise manner by which non-living matter in the cosmos turned into life in the first instance may be a problem that eludes us for generations to come ... (Wickramasinghe, 2010, p. 128)

Proponents of panspermia are already committed to the notion that life is widely distributed in the cosmos (which mostly means 'in deep space') so new scientific observations are turned on their head and become (magically) confirmations of panspermia. For example, when biologically relevant molecules, such as formaldehyde, hydrogen cyanide and glycolaldehyde, are identified in the dust clouds of interstellar space, the most logical interpretation (and the one that makes fewest unsupported assumptions) is that these molecules are evidence of a widespread interstellar chemistry. By contrast, being convinced of the cosmic distribution of bacteria and viruses, the panspermist interprets these molecules as the 'detritus of biology' and deduces that 'interstellar dust particles ... could be largely derived from biology' (both quotations from Wickramasinghe, 2010, p. 120). Consequently, observation of these molecules is seen by the devotee as yet another proof of panspermia. I believe the road to panspermia is paved with uncritical and unscientific conclusions like this. But where does it start? In the pivotal publications (Crick & Orgel, 1973; Hoyle & Wickramasinghe, 1982) panspermia becomes a requisite because the authors believe the difficulties associated with the production of organic molecules on prebiotic Earth and the emergence of a living entity from them are so great that panspermia is the more valid hypothesis:

> Whatever model one chooses the odds against the correct arrangements for the macromolecules of life evolving in any one setting have to be reckoned as being superastronomical, or at least astronomical in measure (Crick & Orgel 1973; Hoyle & Wickramasinghe 1982). It is this inherent difficulty of bridging a vast improbability gap that justifies turning to the wider cosmos for clues... (Wickramasinghe, 2010, p. 120)

This, though, is a creationist explanation, which has similarities with the 'watchmaker argument' (Dawkins, 1986); compare:

(a) The complex workings of a watch necessitate an intelligent designer. As with a watch, the complexity of living organisms and life in general, and the structure of the Solar System and the Universe in general, necessitates a designer. Therefore there must be a god (= designer).
(b) The odds against the correct arrangements for the macromolecules of life evolving on Earth in the time available have to be superastronomical. This inherent difficulty of bridging a vast improbability gap on Earth justifies turning to the wider cosmos as this would give us more time. Therefore the wider cosmos must be full of life.

Magical, creationist and panspermia ideas are intellectually unsatisfactory. What is most interesting and challenging is to understand how life could originate. The panspermia hypothesis appears to give an explanation for the appearance of life on Earth, but only by shifting the problem of the origin of life to some other place at some other time. This is no explanation. As we have seen, some distinguished scientists still think that panspermia is worth defending even against this criticism:

> It has also been argued that 'infective' theories of the origins of terrestrial life should be rejected because they do no more than transfer the problem of origins to another planet. This view is mistaken; the historical facts are important in their own right. For all we know there may be other types of planet on which the origin of life *ab initio* is greatly more probable than on our own. For example, such a planet may possess a mineral, or compound, of crucial catalytic importance, which is rare on Earth... (Crick & Orgel, 1973, p. 341)

Unfortunately, the best alternative to life originating *ab initio* on Earth which is offered conflicts with the Federation's Prime Directive (Starfleet's General Order number 1):

> It now seems unlikely that extraterrestrial living organisms could have reached the earth either as spores driven by the radiation pressure from another star or as living organisms imbedded in a meteorite. As an alternative to these nineteenth-century mechanisms, we have considered Directed Panspermia, the theory that organisms were deliberately transmitted to the earth by intelligent beings on another planet. We conclude that it is possible that life reached the earth in this way, but that the scientific evidence is inadequate at the present time to say anything about the probability... (Crick & Orgel, 1973, abstract)

In other words, on the grounds of the improbability of life originating on Earth from nothing, we offer an alternative explanation, on the probability of which we can say nothing.

And yet, looking at the details of these discussions, I believe the fundamental problem they seek to avoid is unreal. It is a problem of their own making, resulting from a fixed, but erroneous, view of the process they attempt to explain. Proponents of panspermia, in common with many other theorists, seem to have the fixed view that the proposition that 'life evolved on Earth' must mean that the whole Earth was involved, effectively as a single planet-sized reaction vessel supporting the gathering together of all that integrated chemistry as a one-off extraordinary event; life as a singularity. For example, Hoyle & Wickramasinghe (1982, p. 139) include the following:

> Let us make a number of assumptions favourable to the idea that life originated here on the Earth, starting from a broth of organic materials of abiological origin. Let the soup have a volume as great as the whole world ocean and let it have high concentrations of amino acids in particular.

Do we need the 'whole world ocean' to evolve something like a microscopic primitive bacterial cell? Could the process be done in a single drop? Then again, how many drops might there be in a whole world ocean, and how many independent times might the process of 'origin of life' be attempted? And how often? Every hour, every day? For how many millions of years?

I will discuss this question of the scale of the origin of life process in Chapter 7, where I even show that, as a conservative estimate, there were 'ten million more potentially life-generating opportunities in the course of the 840 million years during which life evolved than there are stars in the Universe'. Is that number sufficiently superastronomical for the panspermia believers? Whether it is or not; now, I think, we should put the fanciful notion of panspermia and life on a cosmic scale on the back burner and return to planet Earth.

SIX

ENDOGENOUS SYNTHESIS OF PREBIOTIC ORGANIC COMPOUNDS ON THE YOUNG EARTH

The first serious experimental attempts to make biogenic or prebiotic monomers by adding energy to simple gases were the spark discharge experiments that Stanley L. Miller carried out in Harold C. Urey's laboratory at the University of Chicago. The original experiment consisted of running steam containing a simple gas mixture of hydrogen, ammonia and methane past electrodes supporting a corona spark discharge from an induction coil, then through a steam-condensing loop before emptying back into the boiling flask. The scale of the apparatus was modest; the steam was produced by boiling just 200 ml of water in a 5 l flask and after evacuating air, the apparatus was charged with 10 cm pressure of hydrogen, 20 cm of methane, and 20 cm of ammonia. The boiling and electrical discharge continued for a week;

> ... the water in the flask became noticeably pink after the first day, and by the end of the week the solution was deep red and turbid. Most of the turbidity was due to colloidal silica from the glass. The red color is due to organic compounds adsorbed on the silica. Also present are yellow organic compounds, of which only a small fraction can be extracted with ether ... (Miller, 1953)

The conditions used by Stanley Miller were based on what were then assumed to be realistic conditions on the ancient Earth (Urey, 1952); specifically, that the early Earth would have had a moist and chemically reducing atmosphere of hydrogen, ammonia and methane with a warm liquid ocean (represented by the water in the boiler) and frequent lightning discharges (represented by the spark discharge). For the time the outcome was amazing, for when Miller analysed the solution by two-dimensional paper chromatography run first in butanol/acetic acid/water followed by water-saturated phenol, a ninhydrin spray (the standard way of detecting amino acids) revealed:

> ... glycine, α-alanine and β-alanine are identified. The identification of the aspartic acid and α-amino-n-butyric acid is less certain because the spots are quite weak. The spots marked A and B are unidentified as yet, but may be β- and γ-amino acids. These are the main amino acids present, and others are undoubtedly present but in smaller amounts. It is estimated that the total yield of amino acids was in the milligram range ... (Miller, 1953)

Experiments like this became known as 'Miller–Urey experiments' and they have been repeated many times and all with encouragingly positive results (you can carry out the experiments yourself at this URL: www.ucsd.tv/miller-urey/). Cooking up a reducing gas mixture in a flask exposed to some sort of energy, like electric discharges, or UV light, or X-rays, or just heat, produced a range of biogenic compounds, such as amino acids, sugars, purines and pyrimidines; formed in abundance and in a reasonable time. In the flurry of experimentation that followed Miller's lead it was found that not only amino acids could be made from hydrogen cyanide (HCN) and ammonia (NH_3) in an aqueous solution, but amazing amounts of the purine adenine were formed too. This was a key achievement during the period 1959–1962 (Orgel, 2004), because adenine has great biological significance. It is one of the four heterocyclic bases from which the genetic codes in RNA and DNA are constructed, and it is also a vital component of ATP (adenosine triphosphate). ATP is the universal energy carrier in the cells of all present-day organisms. Because ATP is a universal contributor to living metabolism in the present day it can be assumed to have been a contributor to the most primitive, most original, and most ancient metabolism at the dawn of life.

From their first publication these discoveries created a brisk, and occasionally bitter, debate both within the wider community and within the scientific community. Scientists became very hopeful that questions about the origin of life, and indeed even about the nature of life, would be solved within a decade or so. Stanley Miller's publication in 1953 appeared in the same year as James D. Watson and Francis Crick's revolutionary discovery of the double-helical structure of DNA; and the following decade was the golden age during which molecular biologists established the triplet nature of the genetic code, the meaning of the 64 codons, anticodons, the function of transfer RNA and the essential process of messenger RNA-directed protein synthesis (briefly described in Chapter 8, below). All of this resulted in a rather exultant and self-satisfied view of progress in biology that enabled Sydney Brenner (who was awarded the Nobel Prize in Physiology or Medicine for 2002 in conjunction with Robert Horvitz and John E. Sulston for their discoveries concerning 'genetic regulation of organ development and programmed cell death') to write in 1988:

> In late 1962, Francis Crick and I began a long series of conversations about the next steps to be taken in our research. Both of us felt very strongly that most of the classical problems of molecular biology had been solved and that the future lay in tackling more complex biological problems. I remember that we decided against working on animal viruses, on the structure of ribosomes, on membranes, and other similar trivial problems in molecular biology. I had come to believe that most of molecular biology had become inevitable and that, as I put it in a draft paper, we must move on to other problems of biology which are new, mysterious and exciting. Broadly speaking, the fields which we should now enter are development and the nervous system ... (Quotation taken from the foreword of the Cold Spring Harbor monograph *The Nematode, Caenorhabditis elegans*; Wood, 1988; see also http://nobel-prizewinners.com/brenner/)

The feeling that 'most of the ... problems ... had been solved' by the mid 1960s was probably shared by origin-of-life researchers as more and more Miller–Urey experiments yielded more and more spontaneously synthesised potential prebiotic molecules; leaving only the 'trivial problems' of how all those molecules were assembled into a living organism. But, as with molecular biology, by the mid 1960s many of the most fundamental problems had not even been recognised and the research into life's origins had only just begun.

Fifty years after they were carried out, it became clear that Miller's experiments were even more successful than he knew because the paper chromatography that he used in the 1950s was not sufficiently sensitive to detect more than a few amino acids and was unable to provide reliable quantitative measurement of yields. Stanley Miller died in 2007 and one of his students discovered an archive store containing not only the original equipment from the 1950s' experiment but also sealed vials containing samples of the products from the original experiments.

Twenty-first century analytical techniques are far more sensitive and discriminating, and are able to measure the abundance of a wide range of compounds with great precision. Applying these techniques to the archived samples from the 1950s revealed that the original Miller–Urey experiments had in fact synthesised more than 25 (and possibly as many as 40) different amino acids. Since present-day living organisms use about 20 different amino acids, it is clear that Miller's original experiment was astonishingly successful at synthesising biologically relevant organic molecules from simpler chemicals. It seems perfectly realistic, therefore, to expect that there was plenty of opportunity on the early Earth for such *de novo* synthesis to take place. Indeed, early calculations and experiments indicated that a primeval methane atmosphere would have polymerised so readily under solar ultraviolet radiation alone as to produce an oil slick 1 to 10 metres thick over the Earth in geologically short periods of time (Lasaga, Holland & Dwyer, 1971).

The only criticism that has emerged is that a prebiotic reducing atmosphere was assumed to be the primeval atmosphere by comparison with the outer planets (particularly Jupiter). However, it has become evident that the protoplanet Earth's primeval atmosphere would have been stripped away by the collision with *Theia* and the late heavy bombardment phase of Earth's history. The Moon-forming impact was a key event, not only for the reasons discussed in Chapter 3 but also because the impact itself plainly splits the development of the Earth into two radically different periods: the before-impact and the after-impact. Before the Moon formed, the main events established the size and overall composition of the Earth and probably also established the sizes and compositions of its atmosphere and ocean. The impact itself converted the surface of the Earth into an ocean of magma (molten rock) beneath an atmosphere of rock vapour and steam that displaced the primeval atmospheric gases (see Fig. 3.2 for an outline of the consequences of an impact between Earth and an asteroid only about 20% the size of *Theia*). A consequence of the impact was that all the volatile constituents of the Earth's mantle were driven into the atmosphere and subsequently condensed in sequence. The rock vapour condensed first, followed by non-siliceous minerals and then water. So, after the Moon-forming impact these components were located at the surface and the oceans were probably made of salt water from the start. It is important to stress that the early composition of the atmosphere and the early climate work together to influence the origin of life:

> Apparently, part of the appeal of this reduced atmosphere for the early Earth was that it would have been a greenhouse mixture, keeping the Earth from freezing over during a period when the Sun was dimmer than it is today... (Bernstein, 2006, p. 1690)

I will discuss 'snowball Earth' global glaciations in Chapter 9, but for the present discussion the consensus is that highly reduced methane + ammonia atmospheres are unlikely to have survived the Moon-forming impact even if they were originally present. It is far from certain that the Earth before that impact even possessed such an atmosphere, because it is not clear that the present-day atmosphere of Jupiter is a good guide to the primeval atmosphere of Earth. Certainly, Jupiter has retained its original atmospheric gases because the planet is so massive that the molecules have never escaped its gravitational field. The consequential idea that the Earth also formed with a thick Jupiter-like atmosphere has largely been abandoned because calculations of planetary accretion disc dynamics show that at the time the Earth was formed the temperature in the solar nebula around the orbit of Earth would have driven these less dense gases outwards towards the outer regions of the developing Solar System, and into the region where the gas giants were forming. This would result in the early Earth being devoid of atmosphere (see discussion and references in Bernstein, 2006, p.1690). This is another reason to believe that Earth's atmosphere was formed by outgassing after the Moon-forming impact.

This is the geologically based argument, that Earth's atmosphere accumulated after these events by outgassing from the Earth itself. Add water contributed by occasional cometary impacts and an atmosphere very different from that of Jupiter was formed. Geological data suggest that the Earth has been at or near its current oxidation state for over 3 billion years and probably close to 4 billion. This implies an early atmosphere composed of neutral or perhaps even oxidised gases, such as water vapour, carbon dioxide and sulphur dioxide, though it may have been more neutral (e.g. water vapour + nitrogen + carbon dioxide), but contained essentially no methane or ammonia which are expected to be quickly destroyed by photochemical reactions in the upper atmosphere. In these atmospheres carbon dioxide is the principal greenhouse gas influencing glaciations. Although *de novo* chemical synthesis can still occur in such atmospheres in response to energy input from lightning and/or UV radiation, it is more difficult to make reduced organic compounds like amino acids. There seems to be a consensus that the prebiotic atmosphere of Earth was not a reducing atmosphere; but no consensus yet exists about its exact composition (see discussion in Bernstein, 2006, and Kasting & Howard, 2006). However, Zahnle, Schaefer & Fegley (2010) point out that atmospheres of other bodies that have been generated by impact degassing, like that expected on Earth after the Moon-forming impact, tend to be strongly reducing and volatile-rich. They also make the point that although atmospheres rich in carbon monoxide (CO) or methane (CH_4; which would itself generate hydrogen cyanide and ammonia) are not stable, they are quite likely to have existed several times as lengthy transients following major impacts. But the fact remains that there is no decisive evidence either way about the primitive atmosphere, apart from the general acceptance that oxygen was absent, and consequently there is no agreement yet as to whether gas mixtures of the time were strongly reducing (for example, methane + nitrogen, ammonia + water vapour, or carbon dioxide + hydrogen + nitrogen), neutral (such as carbon dioxide + nitrogen + water vapour), or mildly oxidising (such as water vapour + carbon dioxide + sulphur dioxide). You can get an idea of the current level of argument using this reference as an entry to the debate: Russell (2010).

Before turning to other potential ways of synthesising reduced organic compounds I want to emphasise two points about the Miller–Urey experiments. First, all the starting gases of the range of such experiments that have been carried out have now been found in the interstellar gas clouds; some of the end products of the Miller–Urey experiments are found in space as well, such as formaldehyde and cyanoacetylene (Thaddeus, 2006). Many of the compounds made in the Miller–Urey experiments are known to exist in interplanetary and interstellar space. The Murchison meteorite (Chapter 4) has yielded over 90 amino acids to date, 19 of which are common on Earth. More importantly, the Murchison meteorite demonstrates that the Earth could have received amino acids and other organic compounds by planetary infall. Consequently, if the Miller–Urey experiments fail to account for *de novo* synthesis of reduced organics on Earth, they certainly account for extraterrestrial *de novo* synthesis of reduced organics.

Second, although the general prebiotic atmosphere of early Earth may not have been a chemically reducing atmosphere, the prebiotic Earth was volcanically very active and the gases emerging from volcanic vents were likely to be reducing, moist, hot and turbulent – ideal conditions for lightning and Miller–Urey synthesis, followed by rainfall washout of the products of such syntheses on the warm slopes of the volcanoes where they could escape UV photolysis among the volcanic ejecta and crevices of the lava flows. Consequently, even if the Miller–Urey experiments fail to account for *de novo* synthesis of reduced organics in the general atmosphere of Earth, they certainly account for localised *de novo* synthesis of reduced organics wherever there was a volcanic vent. One of the most interesting of Stanley Miller's experiments of the early 1950s that was reanalysed this century, and one that Miller himself had never published, was an experiment involving conditions similar to those of volcanic eruptions. In this case the experimental apparatus included a nozzle spraying a jet of hot water mist at the spark discharge, simulating a water-vapour-rich volcanic eruption. In 2008, a group of scientists examined 11 vials from Miller's archive left over from this experiment, using high-performance liquid chromatography and mass spectrometry. They found that this volcano-like experiment, which used a gas mixture of hydrogen sulphide + methane + ammonia + carbon dioxide, had produced the most organic molecules of any of the experiments that Miller had performed in the 1950s: 23 amino acids, 4 amines, including 7 organosulphur compounds (including sulphur-containing amino acids) and many hydroxylated molecules, probably formed by hydroxyl radicals produced by steam in the electric discharge (Johnson *et al.*, 2008). This 50-year old experiment was the first synthesis of sulphur-containing amino acids in a spark discharge experiment intended to imitate potential primordial Earth environments. The simulated primordial conditions used by Miller could be a model for ancient volcanic plume chemistry and perhaps offer understanding of the possible roles such plumes could have played in ancient abiotic organic synthesis. Interestingly, the quantitative distribution of the amino acids synthesised in the presence of hydrogen sulphide are very similar to the abundances found in some carbonaceous meteorites (Parker *et al.*, 2011). All of this suggests that hydrogen sulphide may have played an important role in prebiotic reactions in early Solar System environments:

> It is now widely recognized that the first efficient abiotic synthesis of organic compounds under simulated primitive Earth conditions in the context of the origin of life were the classic experiments done by Stanley Miller in the 1950s ... Miller used a reducing gas mixture composed of H_2, H_2O, CH_4, and NH_3, which at the time was believed to be representative of the primitive terrestrial atmosphere ... Many geoscientists today favor an early atmosphere that was likely weakly reducing, containing N_2, CO_2, H_2O, CO and lesser amounts of more reduced species such as H_2S, CH_4, and H_2 ... However, reducing conditions may have been prevalent on the Earth locally or transiently; for example, in the vicinity of volcanic plumes ... and on other solar system bodies (e.g., protosolar nebula, ancient Mars, Titan, etc.). Even with a weakly reducing or neutral atmosphere, recent research indicates that significant yields of amino acids can still be synthesized. (Parker *et al.*, 2011)

There is abundant evidence of major volcanic eruptions for extended periods of time from about 4 billion years ago, which would have released carbon dioxide, nitrogen, hydrogen sulphide (H_2S), and sulphur dioxide (SO_2) into the atmosphere. Experiments using these gases in addition to the ones in the original Miller–Urey experiment have produced more diverse molecules. Also, although more oxidised gas mixtures (where carbon dioxide is abundant) generate few organic compounds of prebiotic interest, recent research shows that spark-experiment yields of ammonia, hydrogen cyanide, and amino acids in carbon dioxide + nitrogen + water vapour mixtures can be more rewarding if the water is made acidic (Zahnle *et al.*, 2010). Nevertheless, during the mid 1990s the widely perceived expectation of poor prospects for the Miller–Urey type of synthesis in what was then expected to be a less-reducing primeval atmosphere led many people to give more thought to what might have been happening in the ocean, and particularly in and around the then recently discovered deep-ocean volcanic hydrothermal systems. Potentially, these provide conditions for the synthesis of organic molecules that might form the basic substrates for the origin of life. In many people's eyes a hydrothermal origin of life is also important because they interpret it as implying that life could be widespread in the Solar System, because hydrothermal systems may exist on many of the satellites in the Solar System.

On Earth in the present day, deep-ocean hydrothermal vents are associated with volcanically active places where tectonic plates are moving apart. First to be discovered were the 'black smokers', observed in 1977 by Jack Corliss and his crewmates on the submersible Alvin in the deep volcanic undersea ridge, called the East Pacific Rise, 2500 m down off the Galapagos Islands (discovery described in Hazen, 2005, chapter 7). Black smokers are large chimney-shaped mounds of material emitting hot geyser water blackened by particles rich in sulphides of lead, cobalt, zinc, copper and silver. This makes them look like smoking chimneys, hence the name 'black smokers'. The vents recycle seawater which has seeped through cracks in the sea floor and is heated by molten rock (magma) deep beneath the ocean floor. The water temperature rises to 350–400 °C, and as it heats up, the water reacts with the rocks in the crust, dissolving minerals and becoming acidic, anoxic and loaded with hydrogen and hydrogen sulphide. Eventually this water emerges as a geyser close to the volcanic rift and the hot hydrothermal fluids rise up through crust carrying the dissolved metals and hydrogen sulphide with them. As they mix with the cold and oxygenated seawater, metal ions in solution combine with hydrogen sulphide to form the black metal sulphides, giving the jet its black smoky appearance, and other minerals come out of solution to construct the 'chimney'. Other types of vent, which are placed further away from the volcanic rift, eject a different composition of minerals; in these cases the hydrothermal fluids mix with seawater under the sea floor so the black sulphides form, and are deposited, beneath the seafloor before the fluid exits the geyser jet. These are 'white smokers' that emit lighter-coloured minerals, such as those containing barium, calcium and silicon (and probably even elemental sulphur). These vents also have lower temperature plumes of alkaline water.

Black smoker submarine geysers of hot water, jetting upward into the cool surrounding ocean, have been promoted as providing an alternative to the spark discharge method of making organic molecules on the early Earth. Reducing gases emerge there today, and as the early Earth was probably even more tectonically active than today, there would have been a great many such vents early on (Fig. 3.1). Laboratory experiments simulating the conditions in hydrothermal vents have yielded *de novo* (abiotic) synthesis of pyruvate, lipids and related compounds, amino acids and peptides, purines and pyrimidines and nucleotide oligomers (discussed by Bernstein, 2006). Thus, the chemistry is extremely promising, but the real driving force behind the belief that deep-ocean hydrothermal vents might be likely sites for the evolution of primeval life is the totally unexpected observation that today these vents, which are abundant along ocean ridges in both the Atlantic and Pacific oceans, feature vibrant and productive ecosystems. These are exotic communities of numerous invertebrate animals (large tubeworms, bivalve molluscs, gastropods, and a range of decapod crustaceans) all ultimately dependent on microbial primary energy producers that are chemical autotrophs at an ocean depth that never sees the light of the Sun.

It was soon suggested that the hydrothermal vents might be the site of the origin of life. Modern microorganisms certainly do thrive in deep hydrothermal ecosystems, and fossil microbes recovered from 3.5 billion-year-old hydrothermal deposits strengthen this suggestion. Another aspect of the argument is that the deep ocean could have offered a much more benign location than the surface at a time when Earth's surface was subjected to heavy meteorite bombardment. Nutrients and chemical energy abound in the deep hot environments of hydrothermal systems, and physical and chemical gradients surrounding them make ideal reactors for abiotic chemical synthesis (Baross & Hoffman, 1985). Several proposals have been made for a sequence of chemical steps that have the potential for rapid emergence of life but the one developed in most detail in recent years is Günter Wächtershäuser's theory of a chemoautotrophic 'iron-sulphur world' origin of life in and on primeval black smokers (Wächtershäuser, 1988, 1992, 2006). Wächtershäuser postulates 'surface metabolists' that arise in an ocean in which the concentration of dissolved organic constituents in the water phase is negligible, effectively zero. That is, their emergence does not depend on a primeval soup containing the organic building blocks for life. Rather than paraphrase such a sharply contrasting notion, it is probably best to quote Wächtershäuser's own description:

> It is proposed here that, at an early stage of evolution, there are precursor organisms drastically different from anything we know. These organisms are acellular and lack a mechanism for division, yet they can grow. They possess neither enzymes nor a mechanism for translation, but they do have an autocatalytic metabolism. They do not have nucleic acids or any other template, yet they have inheritance and selection. Although they can barely be called living, they have a capacity for evolution. Central to the proposed theory is the idea that life at this early stage is autotrophic ... and consists of an autocatalytic metabolism confined to an essentially two dimensional monomolecular organic layer. These surface organisms (surface metabolists) are anionically bonded to positively charged surfaces (e.g., pyrite) at the interface of hot water. The adherence to the positively charged mineral surface is not the result of adsorption (as suggested by Bernal ... in his clay theory) but rather of in situ autotrophic growth of anionic constituents acquiring their surface bonding in statu nascendi ... High-energy phosphoanhydride groups are not required for the formation of covalent bonds. Phosphate groups (whose source is taken to be the mineral substrate) have the sole function of surface bonding. The energy for carbon fixation is provided by the redox process of converting ferrous ions and hydrogen sulphide into pyrite, which is not only a waste product but also provides the all-important binding surface for the organic constituents ... (Wächtershäuser, 1988)

Wächtershäuser's theory is that the entire process from first emergence of 'life' to first emergence of fully integrated cellular life takes place on the iron sulphide (pyrite) surfaces generated by deep-ocean hydrothermal vents (see Fig. 9.1):

> Instead of adsorption, the organism is faced with desorption, that is, a selective detachment of its constituents. This means a negative selection favoring higher anionic bonding strength. Large polyanionic constituents with ever stronger surface bonding are automatically selected: first, polyanionic coenzymes, and eventually nucleic acids and polypeptides. The primitive surface metabolists grow by spreading onto vacant surfaces; they reproduce by reproducing the autocatalytic coenzymes, and they evolve by the environmentally induced ignition of new autocatalytic cycles. The surface metabolists evolve toward higher complexity since the thermodynamic equilibrium in a surface metabolism favors synthesis, not degradation (as would occur in solution) ... (Wächtershäuser, 1988)

The theory makes detailed suggestions about how the first two-dimensional surface metabolists evolve:

> The second stage consists of semicellular organisms still supported by a mineral surface, but with an autotrophically grown lipid membrane and an internal broth of detached constituents. In this stage, a membrane metabolism and a cytosol metabolism appear, first as a supplement to and later as a substitute for the aboriginal surface metabolism. Membrane-bound electron transport chains allow the tapping of other redox energy sources and ultimately of light energy. The cytosol metabolism allows the salvaging of detached constituents and of their chemical energy by catabolic processes and the development of modular modes of synthesis that rely upon energy coupling. Eventually, heterotrophy appears as a by-product of the catabolic salvage pathways. The genetic machinery of the cell develops from surface-metabolic precursors with catalytic imidazole residues glycosidically bonded to a polyhemiacetal backbone of (surface-adhering) phosphotrioses. It produces self-folding enzymes which compete with the mineral surface for bonding the metabolic constituents. In this stage evolution becomes double tracked: an evolution of metabolic pathways and one of the bonding surfaces for their constituents. In the third stage the pyrite support is abandoned and true cellular organisms arise which become free to conquer three-dimensional space. (Wächtershäuser, 1988)

By the early 1990s, the deep–hot hydrothermal origin hypothesis had become widely accepted as a viable, if unsubstantiated, alternative to the Miller–Urey atmospheric chemistry scenario, though not without criticism. Miller & Bada (1988) maintained that the high temperatures of the vent waters (> 350 °C; though the hyperthermophilic bacteria and archaea isolated today from these areas of volcanic activity grow between 80 and 113 °C) would decompose organic compounds rather than lead to their synthesis and would prevent polymerisation of organic compounds. They argued instead that because the ocean waters are cycled through the vents the destructive activity of hydrothermal vents would reduce the concentrations of organic compounds in the oceans of the primitive Earth relative to the land (implying that the vents make the ocean a less likely location for the origin of life). Martin & Russell (2007) pointed out that not all submarine hydrothermal vents are that hot; those known as 'white smokers' release water that is cooler, alkaline and rich in hydrogen (for example, the hydrothermal vents known as the Lost City, near the Mid-Atlantic Ridge at a depth of 700 m, release alkaline (pH 9–11), extremely methane- and hydrogen-rich water at about 40–90 °C). Even before such vents were discovered (in December 2000; Kelley *et al.*, 2001), Russell *et al.* (1994) wrote:

> We propose that life emerged from growing aggregates of iron sulphide bubbles containing alkaline and highly reduced hydrothermal solution. These bubbles were inflated hydrostatically at sulphidic submarine hot springs sited some distance from oceanic spreading centers four billion years ago. The membrane enclosing the bubbles was precipitated in response to contact between the spring waters and the mildly oxidized, acidic and iron-bearing Hadean ocean water. As the gelatinous sulphide bubbles aged and were inflated beyond their strength they budded, producing contiguous daughter bubbles by the precipitation of new membrane. (Russell *et al.*, 1994, abstract p. 231)

This prescient suggestion anticipates that those 'bubbles containing alkaline and highly reduced hydrothermal solution' are released into oceanic water made acidic by dissolved carbon dioxide (see Fig. 9.2). Consequently, within the precipitated gelatinous iron sulphide membrane the hydrothermal solution is relatively depleted of protons whereas the surrounding external oceanic water is relatively rich in protons – a natural proton gradient, making the mineral bubbles released from the alkaline vents naturally chemiosmotic (Russell & Hall, 2002). Furthermore, hydrothermal hydrogen encounters the carbon dioxide dissolved in the ocean and the chemical equilibrium of the hydrogen/carbon dioxide system favours the synthesis of reduced carbon compounds. From these features a complete model for the origin of biochemistry at primeval alkaline hydrothermal vents has been developed:

> The following compounds appear as probable candidates for central involvement in prebiotic chemistry: metal sulphides, formate, carbon monoxide, methyl sulphide, acetate, formyl phosphate, carboxy phosphate, carbamate, carbamoyl phosphate, acetyl thioesters, acetyl phosphate, possibly carbonyl sulphide and eventually pterins. Carbon might

> have entered early metabolism via reactions hardly different from those in the modern [acetyl-coenzyme-A] Wood-Ljungdahl pathway [of carbon dioxide fixation], the pyruvate synthase reaction and the incomplete reverse citric acid cycle. The key energy-rich intermediates were perhaps acetyl thioesters, with acetyl phosphate possibly serving as the universal metabolic energy currency prior to the origin of genes. Nitrogen might have entered metabolism as geochemical NH_3 via two routes: the synthesis of carbamoyl phosphate and reductive transaminations of α-keto acids. Together with intermediates of methyl synthesis, these two routes of nitrogen assimilation would directly supply all intermediates of modern purine and pyrimidine biosynthesis. Thermodynamic considerations related to formyl pterin synthesis suggest that the ability to harness a naturally pre-existing proton gradient at the vent-ocean interface via an ATPase is older than the ability to generate a proton gradient with chemistry that is specified by genes. (Martin & Russell, 2007)

But deep, hot hydrothermal vents are not the only exotic locations for living microbes that have been identified. Microbes were found to thrive in deep buried sediments, oil wells, deep in dry desert sand, even in porous volcanic rocks more than a mile down, apparently favouring mineral surfaces, where interactions between water and chemically unstable rocks provide the chemical energy for life.

> Subsequent drilling studies have revealed that microbes live in every imaginable warm, wet, deep environment – in granite, in basalt on land and basalt under the ocean, in all variety of sediments, and also in metamorphic rocks that have been altered by high temperature and pressure. Anywhere you live, drill a hole down a mile and the chances are you'll find an abundance of microscopic life. (Hazen, 2005)

The discovery that microbial life is widespread deep in the crust of the Earth, just as it is deep in the ocean, gave rise to the notion of the existence of a 'deep, hot biosphere'. Deep, hot life is essentially independent of the surface circumstances, and this is the main reason for favouring these sites as locations for the origin of life, as this independence offers protection from meteorite or cometary impacts and from high UV and other radiation fluxes from the Sun. Gold (1992) suggested that, in the present day, the deep, hot biosphere 'in mass and volume ... may be comparable with all surface life.' And not only 'hot', because microbes survive under miles of Antarctic ice; inevitably, deep cold-water locations, under thick ice layers, have also been suggested as locations for the origin of life. Providing some support for this argument is the fact that chemical syntheses have been shown to proceed quite well under low temperature conditions:

> In what must hold the record for the longest experiment ever performed, Stanley Miller and co-workers published a report on the prebiotic

> synthesis of purine and pyrimidine bases from a dilute solution of ammonium cyanide solution frozen at −78°C for 27 years! ... Similarly, however, less dramatically, it had been shown previously that hydroxy, amino and carboxylic acids can be formed in aqueous solutions at −10°C ... Not only can starting materials be made under these conditions, but template-directed RNA oligomerization was recently shown to be remarkably effective in frozen seawater with temperature variation, yielding chain lengths of up to 400! ... (Bernstein, 2006, and references therein)

The widespread distribution of microbes at great depth in the crust today certainly demonstrates the ability of living organisms to adapt to challenging environments, but I am not convinced that this has much relevance to the origin of life. Rather, I follow Bernstein (2006): 'Incidentally, it seems to me that the conditions under which life first developed might well be more restricted than those under which life can survive or thrive thereafter', though I would add the suggestion that the conditions under which life first developed could have been more benign than those to which life adapted subsequently. In addition, suggesting that life originated somewhere deep within the crust of the Earth, hot or cold, seems to carry with it the requirement that first life was autotrophic with its energy supply coming from chemical sources in fluids that migrate upward from deeper levels in the Earth. Yet I have gone to great lengths so far (and have not finished yet; see Chapter 7) to emphasise that the early Earth was, like the rest of the Universe today, full of organic compounds that could serve as nutrients for any organism that could digest them. I believe, therefore, that the first organisms were more likely to be heterotrophs than autotrophs. In present-day biology a heterotroph is an organism that cannot 'fix' carbon; that is, it cannot reduce carbon dioxide from the atmosphere to make carbohydrates, organic acids, fats and proteins. Heterotrophs use already reduced (organic) carbon compounds that they take from their environment to support their growth and development. This contrasts with autotrophs, which can use an external energy source to fix carbon.

In the present day autotrophs are represented by photoautotrophs, such as plants, algae and photosynthetic bacteria, which use energy from sunlight, and lithoautotrophs (bacteria) that use reduced inorganic chemical compounds to produce their organic compounds from inorganic environmental carbon dioxide. In the present day most heterotrophs are saprotrophs (like most fungi) that feed on the dead organic remains of other organisms. At the origin of life the first heterotrophs would have had available to them all the organic compounds accumulated on Earth from non-biological synthesis (Chapter 7). Interestingly, in the present day the heterotrophs include groups of organisms described as photoheterotrophs (most purple and green bacteria) that use light to make adenosine triphosphate (ATP), and chemoheterotrophs that produce ATP by oxidising inorganic chemicals. In neither case is the external source of energy used directly to fix carbon. But they may indicate a potential evolutionary pathway by which autotrophic primary producers might have emerged: from heterotroph without external energy input, to heterotroph plus supplemental energy input through (for example) anoxygenic photosynthesis, to autotroph with (for example) oxygenic, photolysing, photosynthesis and no dependence on external supplies of reduced carbon compounds.

In the next three chapters I am going to discuss the recipe for life by collecting together the various potential sources of prebiotic organic compounds, I will then attempt a brief definition of what we mean by 'life', so we can recognise it when we see it, and I will discuss in Chapter 9 the transition between not alive and alive and outline the various models put forward to account for this.

All of these topics have been discussed by others many times before, and I build on their erudition and add my own slice of mycological understanding. To make it clear what prompted my own thoughts I will end this chapter with a quotation from Robert Hazen (2005), which is taken from a section he entitles 'Three scenarios for the origin of life':

> The greatest mystery of life's origin lies in the unknown transition from a more-or-less static geochemical world with lots of interesting organic molecules to an evolving biochemical world in which collections of molecules self-replicate, compete, and thereby evolve. How that transition occurred seems to boil down to a choice among three possible scenarios.
>
> 1. Life began with metabolism, and genetic molecules were incorporated later: following Gunter Wächtershäuser's hypothesis, life began autotrophically. Life's first building blocks were the simplest of molecules, while minerals provided chemical energy. In this scenario, a self-replicating chemical cycle akin to the reverse citric acid cycle became established on a mineral surface (perhaps coated with a protective lipid layer). All subsequent chemical complexities, including genetic mechanisms and encapsulation into a cell-like structure, emerged through natural selection, as variants of the cycle competed for resources and the system became more efficient and more complex. In this version, life first emerged as an evolving chemical coating on rocks . . .
> 2. Life began with self-replicating genetic molecules, and metabolism was incorporated later: according to the RNA World hypothesis, life began heterotrophically and relied on an abundance of molecules already present in the environment. Organic molecules in the prebiotic soup, perhaps aided by clays or PAHs or some other template, self-organized into information-rich polymers. Eventually, one of these polymers (possibly surrounded by a lipid membrane) acquired the

ability to self-replicate. All subsequent chemical complexities, including metabolic cycles, arose through natural selection, as variants of the genetic polymer became more efficient at self-replication. In this version, life first emerged as an evolving polymer with a functional genetic sequence ...

3. Life began as a co-operative chemical phenomenon arising between metabolism and genetics: a third scenario rests on the possibility that neither protometabolic cycles (which lack the means of faithful self-replication) nor protogenetic molecules (which are not very stable and lack a reliable source of chemical energy) could have progressed far by themselves. If, however, a crudely self-replicating genetic molecule became attached to a crudely functioning surface-bound metabolic coating, then a kind of co-operative chemistry might have kicked in. The genetic molecule might have used chemical energy produced by metabolites to make copies of itself, while protecting itself by binding to the surface. Any subsequent variations of the genetic molecules that fortuitously offered protection for themselves or for the metabolites, or improved the chemical efficiency of the system, would have been preserved preferentially. Gradually, both the genetic and metabolic components would have become more efficient and more interdependent.

Such a 'dual origins' model might at first seem to introduce a needless complication (not to mention sounding like a wishy-washy compromise). Nevertheless, exactly this kind of symbiotic coupling of metabolism and genetics is now thought to have occurred early in the history of cellular life. Crucial features of our own cells suggest an ancient co-operative merging of early, more primitive cells. If experiments establish easy synthetic pathways to both a simple metabolic cycle and to an RNA-like genetic polymer, then such a symbiosis may provide the most attractive origin scenario of all (Hazen, 2005, chapter 19).

SEVEN

COOKING THE RECIPE FOR LIFE

As I have shown above, there are several ways to synthesise prebiotic organic compounds, all of which seem to be realistic, though the extent of their contribution to the early Earth will in most cases depend on the exact environmental circumstances in the place and at the time that they arise. Nevertheless, any one or (most likely) all of the following processes will allow for the synthesis of at least some organic molecules that would contribute to making the Earth habitable (Bernstein, 2006; Cady, 2001; Ehrenfreund *et al.*, 2002, 2005; Ehrenfreund & Cami, 2010; Zahnle *et al.*, 2007, 2010):

Spark discharge synthesis: in practice this means lightning but also includes the effects of high energy solar radiation, especially UV, in the atmosphere. The outcome depends on the oxidation state of the atmosphere and on how much hydrogen is/was present; but although yields of organic products might be limited in the general atmosphere, considerable quantities of biogenic organic molecules could be made locally during volcanic eruptions (Parker *et al.*, 2011).

Hydrothermal vent synthesis: high temperature chemistry in the water outflow of deep-ocean (black smoker) vents can generate interesting chemical pathways but the stability of amino acids and other compounds in these environments remains problematic, and such vents may actually purge the seawater of biogenic molecules. Cooler, alkaline (white smoker) vents support a wide range of organic synthesis and readily support production of bubbles bounded by an inorganic membrane across which a proton gradient is naturally established (Martin & Russell, 2007; Simoncini, Russell & Kleidon, 2011), which might be the primeval ancestor of chemiosmotic coupling; 'an energy source that must have been available to emergent life' (Russell, 2010).

Extraterrestrial input: ultimately all the elements on Earth date back to the previous generation of stars. Many other chemical structures were synthesised in the interstellar medium, or in the solar nebula, or subsequently in the Solar System. Since it first began to form, the Earth has received a lavish collection of organic molecules from the space that surrounds it. For the whole of its life, and up to the present day, tons of organic molecules synthesised in interstellar and interplanetary space are swept up by the Earth every day, delivered to the Earth by asteroids, comets and smaller fragments such as meteorites and interplanetary dust particles. Interestingly, even the impact of meteorites would create new molecules. Ballistic impact shock experiments with aqueous solutions of amino acids show that a large fraction of the amino acids survive the impact, and the impact process resulted in the formation of peptide bonds, to form amino acid dimers and cyclic diketopiperazines (Blank *et al.*, 2001). Furthermore, the impact craters themselves provide habitats because the impact-shocked rocks have many times more pore spaces than intact rock. These provide microhabitats in the near-surface environment of the rocks, which are moisture-retaining and UV-protected microenvironments that might have acted as sites for the concentration of reactants for prebiotic syntheses (Cockell, 2004).

It is most likely that a combination of all these sources contributed to the building blocks of life on the early Earth. The quantitative contributions of organic matter to the prebiotic environment by these different sources have been estimated by Bernstein (2006; see his figure 8, p. 1696). I have already mentioned above (Chapter 4) the estimated input of 10^9 to 10^{13} kg of carbon per year from extraterrestrial sources to the Earth during the 600-million-year late heavy bombardment phase (an overall total of a minimum of 6×10^{12} metric tonnes). Bernstein estimates that Miller–Urey-type synthesis of organic matter would be in the same range as what might have come from comets (Chyba & Sagan, 1992), that is, the equivalent of 10^7–10^9 kg yr^{-1}; and that hydrothermal vents might have contributed $1–3 \times 10^8$ kg yr^{-1}. It seems to me that with these quantities of ready-made organic compounds being available every year on Earth the first organisms (or, indeed, the first pre-alive organisms – the ones that have not yet quite made the grade of being 'alive') did not need to be autotrophs; they do not need to start by being primary producers. Everything they need to start is around them; they need only to be heterotrophs. With regard to the availability of abiotically synthesised organic compounds, Bernstein concludes:

> ... there were a number of different sources of complex organics on the early Earth making all kinds of interesting organic molecules. Clearly,

> these different organic syntheses contributed different kinds of compounds depending on the conditions. However, it does seem inevitable that certain kinds of molecules, such as simple amino acids and fatty acids, must have been present because we see them arise repeatedly from divergent settings like meteorites, spark discharges and hydrothermal vents. (Bernstein, 2006, p. 1696)

Bernstein (2006, p. 1694) also states that these 'are exactly the kinds of species that Darwin, Oparin and Haldane would have been happy to use to flavour the primordial soup, and they come ready made to the Earth' but he also offers a 'new gustatory metaphor to replace the primordial soup' in the form of a 'boxed cake mix that requires the addition of some moisture and fat.' Bernstein's food preparation instructions are as follows:

> 1 turn on star, preheat solar nebula
> 2 mix gas, dust, and rock, add radiation and initiate aggregation
> 3 cook for 100 million years until planet forms.
>
> Prepare additional mars sized body for collision to make moon. (Bernstein, 2006; see his figure 4, and note the credits to NASA Ames and Karen Harpp of Colgate)

The off-the-shelf cake mix is offered as a way to think of what was needed for life to arise on Earth. 'The idea is that unlike a soup from a can that comes containing everything and needs only to be warmed up, a cake mix has most of what you need but usually requires that something be added, such as some milk or butter.' By analogy, the early Earth acquired water and organic matter from asteroids and comets, which it is thought made life possible on a habitable planet.

Neither cake mix nor soup is a completely satisfactory metaphor and it's all a matter of completeness, concentration and scale:

(a) cake mix implies a localised complete mixture of dry ingredients, with concentration controlled by addition of water;
(b) soup implies a homogeneous mixture of all ingredients, which are already at the right concentration.

And yet we are postulating a chemical evolution, a gradual emergence of a living entity so there is neither likelihood nor necessity for all the ingredients of the final entity to be present at the beginning. Indeed there may have been several beginnings of what I will call pre-alive systems in different locations, each reaching a point of equilibrium appropriate to the local set of ingredients until encountering another complementary pre-alive system, derived from some other location, with which a mutual co-operative arrangement can be established to produce a more advanced, more integrated pre-alive system. Which might then encounter a third complementary pre-alive system with which it might compete and/or establish a mutual co-operation; and so on, and so on.

Talk of different locations brings up the question of scale and ways in which pre-alive systems formed in different locations might come into contact. Remember that the designation of the primordial 'soup' originates from Haldane's description of the prebiotic ocean as 'a hot thin soup' (Haldane, 1929); but why think in terms of an ocean? We are theorising about the origin of a pre-alive system that is a submicroscopic organised arrangement of molecules (emerging 'from a random mix of essential but disordered components' according to Fenchel, 2002); this could happen in a single drop of water; it doesn't need an ocean, or even anything as large as a pond. So I believe we should think in terms of small quantities of water; certainly not exceeding the 200 ml of water that Stanley L. Miller used in his first spark discharge experiments, and probably very much less.

On every ocean there would be storms, and from every storm-tossed wave, spindrift, the spray blown from cresting waves that drifts as an aerosol into the atmosphere. An aerosol is a suspension of fine solid particles or liquid droplets in a gas; commonly encountered examples are clouds and smoke, but aerosols are ubiquitous in our atmosphere and even clear air may contain tens of millions of solid particles and liquid droplets [view http://earthobservatory.nasa.gov/Features/Aerosols/]. Aerosols drift throughout the Earth's present-day atmosphere from the stratosphere to the surface. They range in size from a few nanometres diameter (1 nm = 10^{-9} m), on a par with the size of the smallest viruses, to several tens of micrometres (1 μm = 10^{-6} m), about the size of a pollen grain or fungal spore. Atmospheric aerosols have been very seriously suggested as prebiotic chemical reactors; because those of the present day seem to be ideal:

> Aerosol particles in the atmosphere have recently been found to contain a large number of chemical elements and a high content of organic material. The latter property is explicable by an inverted micelle model. The aerosol sizes with significant atmospheric lifetimes are the same as those of single-celled organisms, and they are predicted by the interplay of aerodynamic drag, surface tension, and gravity. We propose that large populations of such aerosols could have afforded an environment, by means of their ability to concentrate molecules in a wide variety of physical conditions, for key chemical transformations in the prebiotic world. We also suggest that aerosols could have been precursors to life, since it is generally agreed that the common ancestor of terrestrial life was a single-celled organism. (Dobson *et al.*, 2000)

Today, aerosols have major impacts on our climate and health; about 10% (usually the most toxic) result from human actions, such as:

(a) industrial activities (may produce toxic gases, sulphates, organic carbon, black carbon, nitrates);
(b) burning fossil fuels (produces large amounts of sulphur dioxide, which reacts with atmospheric water vapour to create sulphurous acid aerosols, and also adds to the greenhouse gas content of the atmosphere);
(c) land clearance by deliberate biomass burning, which produces smoke of organic carbon and black carbon (soot).

About 90% (by mass) of current aerosols have natural origins:

(a) wind-driven spray from ocean waves flings sea salt aloft as one of the most common types of aerosol;
(b) on land, sandstorms and duststorms blow small pieces of mineral dust from deserts, and from drought-hit agricultural fields, into the atmosphere, as the second most common type of aerosol in the present day;
(c) volcanic ash in the present atmosphere is about the third most common type of aerosol.

On the primeval Earth these last three sources are likely also to have been present, although probably in a different ranking of importance. It is generally considered that the earliest oceans are likely to have covered most of the planet; landmasses were few and until continent-sized dry lands were produced (by volcanic activity), sandstorms and duststorms were unlikely to make much contribution. In the beginning, therefore, the sea, the volcanoes, and the atmosphere were the principal sources of aerosols.

Large populations of aerosol droplets would have provided an environment for the concentration of prebiotic organic molecules and for them to be chemically transformed by reaction together and by their exposure to fluctuation in humidity, temperature, and sunlight exposure at different altitudes and latitudes in the atmosphere. In the present day:

> ... measurements of individual aerosol particles in the upper atmosphere show that, on average, organic molecules comprise roughly 50% of the mass of the upper tropospheric aerosol particles in the tropics ... These measurements ... also show that a total of 45 elements have been detected in the atmospheric aerosol at altitudes between 5 and 19 km. All of the elements essential for present-day life are there, including much higher concentrations of carbon, nitrogen, and trace elements than are found in seawater ... (Dobson *et al.*, 2000)

Much of the organic content of atmospheric aerosols is located at their surface; this is interpreted to mean that aerosol particles are coated with organic surfactants, such as long-chain carboxylic acids ('fatty' acids), that self-organise at the air–liquid interface with their hydrophobic hydrocarbon tails exposed to the atmosphere and their polar heads inserted into the aqueous core. Dobson *et al.* (2000) also suggest that the diversity of organic

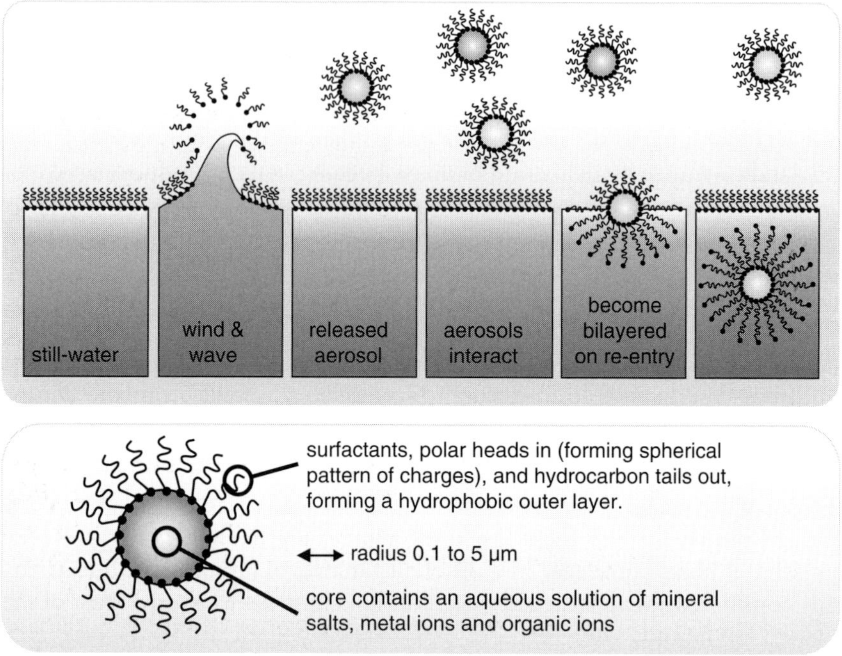

Fig. 7.1 The cycle of aerosol formation, reaction and return to water. The first panel shows organic surfactants, with their polar groups in the water and their hydrocarbon chains in the air concentrated at the water surface (other more soluble organic molecules will also tend to partition to the air–water interface). Aerosols are formed by wind-driven wave action where formation of an aerosol droplet with a partial monolayer of surfactant occurs at the crest of the wave. Released aerosols can drift in the atmosphere for some time and over large distances. Evaporation of water from the interior while airborne produces a complete monolayer of surfactant; exposure to sunlight, lightning and temperature fluctuations in atmospheric suspension stimulates chemical reactions within the aqueous droplet and between contents of the droplet and molecules, ions and radicals from the atmosphere. Aerosols from ocean waves (spindrift) can interact (and merge) with aerosols from other sources that contain different chemicals (volcanic plumes and jets from both submerged and exposed volcanoes). Eventually, the aerosol droplet will settle towards the water surface, shown in the final two panels, where it will interact with the resident layer of organic surfactants and acquire a second layer of surfactant that permits re-entry into the water as a stable vesicle with a bilayered surface membrane. The molecules are greatly exaggerated in size for graphical convenience. (Based on figure 2 in Dobson *et al.*, 2000.)

aerosols would have been increased by potentially repeated cycles of division and coalescence of and between aerosol droplets and particles. Figure 7.1 shows this cycle of aerosol formation, reaction and interaction while suspended in the atmosphere and eventual return to water as a stable vesicle with a bilayered membrane, and although developed from observations of present-day atmospheric aerosols the cycle depends on fundamental physical properties that are just as true of ancient times as they are for present times.

Experiments have shown that marine aerosols of the present day can have an exterior film of amphiphilic fatty acids (that is, a lipid membrane) and that such aerosols are capable of a form of division (Donaldson, Tuck & Vaida, 2002; Donaldson *et al.*, 2004). Other experiments show that subjecting aqueous aerosols containing ferrous and other ions to spark discharges in methane + nitrogen + hydrogen atmospheres yields amino acids, carboxylic acids, heterocyclic compounds and ferric ferrocyanide, the last of which might be considered 'concentrated cyanide' that could make hydrogen cyanide (HCN) available for polymerisation into further organic compounds for prebiotic synthesis (Ruiz-Bermejo *et al.*, 2007). That is, the aerosol droplets are not only reaction vessels for chemical reactions between preformed organic molecules already present in the water from which the aerosol was formed, but the aerosols also serve as reaction vessels for the synthesis of entirely new organic molecules.

There was no shortage of sources of aerosols on the early Earth because the early Earth was, quite literally, a chaotic place. The planet rotated more rapidly than today, resulting in:

(a) more rapid day/night temperature changes;
(b) greater Coriolis and centrifugal forces, meaning more extreme rotation of winds and currents of the air in the atmosphere or the water in the ocean.

In addition:

(a) the Moon was closer than it is today (producing enhanced and more rapidly changing lunar tidal forces);
(b) the atmosphere was denser than it is today (greater mass per unit volume, so greater inertia, meaning stronger winds);
(c) the ocean was warmer, more extensive and shallower than today (resulting in more energetic circulations and currents);
(d) and there was more volcanism than there is today, producing scattered emergent volcanic island landmasses with which the ocean swells and waves would collide after thousands of miles of uninterrupted flow, and on whose slopes cyclonic storm clouds would rain torrentially after thousands of miles of raging across the ocean.

All of this adds up to turbulence in atmosphere and ocean on a massive scale, to which was contributed explosive volcanic ejections of dust, steam and sulphurous gases, both on the land represented by those volcanic island landmasses and in the shallows and deeps of the ocean as magma jetted into overlying water. So, from every source, billions upon trillions of aerosol particles were produced, storm after storm, year after year; and every one of those aerosol particles was a potential independent experiment in creating life or its component pre-alive systems. Donaldson *et al.* (2002) have even attempted to calculate how many aerosol droplets there might have been. By estimating the length of time between the last of the sterilising impacts and the earliest geological evidence for life they estimate the time interval during which life originated to be 840 million years. During this period, they estimate there would have been up to about 10^{14} aerosol cycles between the water surface and the atmosphere (roughly speaking that means about 100 000 storm-lashed waves each year); and at any one time the atmosphere (from the surface to an altitude of 16 km) would contain roughly 10^{17} aerosol droplets containing appropriate reactive molecules (as the surface area of the Earth is just over 5×10^8 km^2, this corresponds to about 200 million aerosol droplets suspended in the column of atmosphere over each square kilometre of surface). In total, therefore, Donaldson *et al.* (2002) estimate the number of life-generating opportunities in aerosol reaction-vessels to be of the order of 10^{31}. I personally think this is a gross underestimate (there must be many more aerosol droplets in the atmosphere at any one time, and there must be more storm-tossed waves in a year), but it doesn't matter. The fact is that 10^{31}, even as a minimum estimate, is a very large number. If you Google 'stars in the universe how many' you'll get values of between 10^{22} and 10^{24}; so at the very least, there were ten million more potentially life-generating opportunities in the course of the 840 million years during which life evolved than there are stars in the Universe.

Not all of these experiments would have had positive outcomes (only one needs to work, after all). In many atmospheric aerosol droplets prebiotic organic molecules would be photolysed by excessive UV radiation. Many others might lose their water by evaporation causing exotic chemical changes in the concentrating solution and eventual solidification. Still more might freeze in the high atmosphere, which would again change the direction of their chemistry. Yet the products of all these failures would eventually be rained out of the atmosphere to be returned to the ocean or land surface and given the chance of once again being returned to the aerosol reactors with the next crashing wave. So it would be a rare aerosol droplet, perhaps, with just the right initial combination of chemicals exposed to just the right combination of heat and radiation for the reactants to form some pre-alive system, which was capable of sustaining itself within the aerial droplet and eventually in the liquid film in which it was deposited.

The greatest puzzle of all time is how life arose; yet the second-greatest puzzle is how it arose so quickly. In this story of the origin and emergence of life on Earth we are used to thinking in terms of billions of years, but life originated in much less time than this. Conditions during the late heavy bombardment phase were far too harsh for any steps that might have been taken on the surface towards life during a period of calm to survive the sterilising effects of the next sequence of impacts. Life at the surface could arise and be sustained only after the conclusion of the late heavy bombardment. This is dated to about 3.8 billion years ago, at which time the surface of the Earth stabilised to form the natural habitat suitable for the origin of life.

Of course, the natural habitat may not have been at the surface; deep marine hydrothermal sites or deep crustal sites may have been sufficiently protected from some of the later, lesser impacts for extreme thermophilic species to survive (and give rise to the now existing major groups, Bacteria and Archaea). Gogarten-Boekels, Hilario & Gogarten (1995) interpret molecular phylogenies this way; suggesting that life emerged early in Earth's history, even before the time of the heavy bombardment was over, so that early life forms had colonised already those extreme habitats that allowed at least two prokaryotic lines to survive a late 'nearly ocean boiling impact, though they do stress that this should not be interpreted as evidence that life originated in extremely hot environments. However, not even the most extreme thermophiles could withstand the impacts, dated around 3.9 billion years ago (for which evidence is preserved on the Moon) that generated enough heat to turn the Earth's rocky crust to an ocean of magma and its atmosphere to vaporised rock.

> We find that if the deep marine hydrothermal setting provided a suitable site, abiogenesis could have happened as early as 4000 to 4200 Myr ago [million years ago], whereas at the surface of the Earth abiogenesis could have occurred between 3700 and 4000 Myr. As a result of the lack of a Hadean terrestrial rock record, many of the constraints on the mode, environment and timing of abiogenesis are derived from laboratory simulations or from theoretical extrapolations to early terrestrial conditions, either forward in time from the formation of the Solar System or back from the Archean record ... (Maher & Stevenson, 1988)

Compare this with the fact that sediments in the northwest of Western Australia that are 3.5×10^9 years old contain very convincing cellularly preserved filamentous microfossils (Derenne *et al.*, 2008; Schopf, 1993). These are among the oldest fossils known but their morphometrics are consistent with modern filamentous cyanobacteria (Boal & Ng, 2010). On the basis of this evidence, therefore, fully differentiated, photosynthetic bacteria were established on the Earth's surface within a few hundred million years of the end of the late heavy bombardment.

> The timescale for abiogenesis under these different circumstances is not known, but is potentially quite short ... We have chosen plausible values of 10^5 to 10^6 years for the hydrothermal model, 3×10^5 to 3×10^6 for the warm puddle case, and 10^6 to 10^7 for the soil hypothesis.

> We acknowledge that these are highly uncertain, but even large changes in these values will not substantially alter the results. (Baltscheffsky *et al.*, 1997)

So these calculations estimate that the first experiments with 'life' finally emerged and matured over a period of between one hundred thousand and ten million years. This compares with the 840 million years used by Donaldson *et al.* (2002) in their estimates about aerosols shown above. If the shorter time estimate (10^5 years) for the period during which life evolved is more realistic it might mean there were 'only' 10^{27} life-generating opportunities in aerosol reaction-vessels; though this is still a thousand times more than there are stars in the Universe!

It may even have taken several starts before life surpassed the less than ideal conditions at the surface (see Fig. 3.1). What is certain is that once life emerged, it learned to adapt quickly, taking advantage of every available refuge and energy source (e.g. photosynthesis and chemosynthesis), an attribute that eventually led to complex metabolic life and even our own existence. Perhaps we should pause here and try to define, in Chapter 8, what we mean by 'life', so we can recognise it if and when we see it.

EIGHT

'IT'S LIFE, JIM ...'

The definition of life is an old philosophical problem, which has been made even more complex by the advent of the current understanding of nucleic acids and the impact of this on genetics, and the ascendancy of molecular biology. (For careful, detailed and entertaining discussions of the topic 'what is life' I would recommend the following references: Benner, 2010; Davies, 2006; Dyson, 1999; Fenchel, 2002; Gánti, 2003; Hazen, 2005; Koshland, 2002; Luisi, 2006; Lurquin, 2003.) But the most significant quotation, and the most important to place right at the start of this chapter is this:

> What is the definition of life? I remember a conference of the scientific elite that sought to answer that question. Is an enzyme alive? Is a virus alive? Is a cell alive? After many hours of launching promising balloons that defined life in a sentence, followed by equally conclusive punctures of these balloons, a solution seemed at hand: 'The ability to reproduce – that is the essential characteristic of life,' said one statesman of science. Everyone nodded in agreement that the essential of a life was the ability to reproduce, until one small voice was heard. 'Then one rabbit is dead. Two rabbits – a male and female – are alive but either one alone is dead.' At that point, we all became convinced that although everyone knows what life is there is no simple definition of life ... (Koshland, 2002)

I put high significance on this particular quotation for the phrases 'everyone knows what life is' and 'there is no simple definition of life'; these are warnings of the nature of the road ahead if you choose to venture down this path. It is also evident that most authors are concerned to present (and bring to prominence) their own uncompromising opinion: Hazen (2005) writes:

> It's amazing how the 'What is life?' question sparks arguments and fosters hard-line positions. Scientists excel at many things, but compromise is not always one of them ... (Hazen, 2005, chapter 2, p. 27)

Hazen notes that there are two complementary approaches to research on the origin of life and the definition of life, which he characterises as 'top-down' versus 'bottom-up'. The top-down approach starts with living and fossil organisms that can be found on present-day Earth and uses these catalogues to identify the most primitive and the most highly adapted and the evolutionary trends that gave raise to present-day organisms. Those interested in the origin of life will inevitably find the most relevant clues about life's early chemistry in primitive microbes and ancient microfossils. But Hazen (2005, p. 26) points out that the statistical flaw in using this approach to establish a definition of life is that it is inevitably limited to life on this planet. No matter how large the catalogues of organisms and fossils might become, together they all constitute just a single observation of a planet filled with life based on biochemically sophisticated cells containing DNA and proteins. We don't know if any other form of life is possible. Our definition of life is not enlightened and top-down research is correspondingly limited.

In contrast, those interested in the bottom-up approach attempt to mimic the chemistry of ancient Earth environments with the eventual aim of creating a 'living' chemical system in the laboratory. Hazen (2005) highlights that this approach leads to a range of definitions of what is alive, because each researcher tends to shape the definition in terms of their own chosen specialty:

> One group will focus on the origin of cell membranes; to them, life began when the first encapsulating membrane appeared. Another team studies the emergence of metabolic cycles, so naturally for them the origin of life coincided with the origin of metabolism. Still other groups investigate primordial RNA ..., viruses, or even artificial intelligence, and each group hawks its own definition of life's first appearance ... (Hazen, 2005, p. 27)

Here's another warning, then: beware of passionately held opinions, always seek out alternatives.

I want to hark back to the Koshland (2002) quotation above to add the additional warning that everyone might know what life is, but they are not all thinking about the same thing. I have read all the reference sources I've cited above, and a great many others as well, and I would claim that four different kinds of 'life' are discernible in the thought processes of the different authors writing on the origin of life:

(a) molecular/cellular life; the primary topic for microbiologists, bacteriologists and those primarily or even solely interested in the transition between chemistry and biology; the early (molecular) stages in these discussions include the coacervates, lipid vesicles, chemotons, surface metabolists, chemiosmotic vesicles and pre-alive systems discussed above, as well as the iron-sulphur world, protein world, or RNA world;
(b) multicellular life; very often limited, explicitly or implicitly, to 'life at the metazoan level, i.e. the life of those "biological supersystems" whose elementary units are living cells' (Gánti, 2003, chapter 1), and likely to confuse the origin of life discussion with concern for embryology, pattern formation and other developmental phenomena;
(c) intelligent life; discussion is often concentrated on animal behaviour, ultimately, human behaviour and the emergence of consciousness and civilisation;
(d) alien intelligent life; those who seek an answer to the question 'are we alone in the Universe' express little kinship with the bacterial grade of organisation; they mostly expect an intelligent conversation with ET, usually as an exchange of binary mathematics, and therefore look for definitions based on scientific knowledge and civilisation. Dedicated exobiologists are more likely to be interested in the extraterrestrial evolution of bacteria, it's true, but their criteria for a definition of life hinge around 'biosignatures', which are essentially some forensic evidence of life's existence that is open to detection and measurement by long-range instrumentation.

And in all of these strands of discussion it is often possible to identify the animal centricity, to which I refer in Chapter 1. Even at the science fiction boundary, where computer simulations of imagined life on other worlds are created, it is remarkable how quickly those modelling extraterrestrial evolution limit the horizons of their imagination to animal life.

Nobody seems to ask the question 'are animals necessary?' Inevitably, I suppose, since the organism asking the question is an animal and nobody likes to be shown to be logically redundant; but if you were designing and building the Earth's biosphere anew what components would you need? In outline, my answers are: bacteria would be necessary to get the whole thing started; photosynthetic organisms (bacterial to begin, eventually plants) would be essential as primary producers, and a type of organism (bacteria to begin, eventually fungi) able to externally digest waste materials would be essential as primary recyclers and enablers for the primary producers. But animals? Animals only eat the rest of the biosphere and procreate; they make no constructive contribution to the biosphere (except when plants and fungi manipulate and make use of them). The Earth's biosphere would not exist without bacteria; would not exist without fungi, and would not exist without plants. But the Earth's biosphere would be perfectly balanced and healthy overall if animals had never evolved. Of course, since they could evolve, they did evolve and to their credit they contribute their own spectrum of biodiversity; but they are not essential to life on Earth. It is arguable that animals are the pinnacle of the expression of life on Earth, but remember it would be an animal doing the arguing, so it would be difficult to find a trustworthy balance in the argument. You may detect an antipathy towards animals here and in Chapter 1 where I first raise this complaint, but it is more a plea for that balance than an antagonism to any one kingdom. I do not advocate that we purge animals from planet Earth. For one thing this exercise would bring my writing career to an abrupt end. But I do advocate that the animals that write, film, video and think about life on Earth should extend their imaginations beyond the limits of their own kingdom to encompass the other kingdoms and domains that share the planet with them.

Although philosophers and theologians often introduce more intangible viewpoints on the definition of life debate, their contributions are not universally appreciated. Benner (2010) tells a personal story, from which he concludes 'scientists ... find little useful in these comments':

> I consulted the local philosopher of science who sent back the message that 'philosophers are weary of satisfactory definitions of any non-trivial term these days.' He continued: 'According to the classical philosophical understanding of definition, a definition must give both necessary and sufficient conditions, and must do that as a matter of the meaning of the term ... A definition, on this classical understanding, must be *a priori* – at least its justification must be *a priori* (because it is supposed to be an analytic claim – true solely in virtue of the meaning of the terms involved). It turns out that, when understood this way, [a definition] is almost impossible to find.' Little wonder that scientists attempting to create life in the laboratory or launch an instrument to Mars find little useful in these comments ... (Benner, 2010, p. 1029)

In this quotation Benner (2010) clearly makes the point that defining life is particularly important to exobiologists and his article concentrates on the various definitions of life held in the astrobiology community. For this community it is important to be able to have a definition of life that can be rephrased to allow prediction of the sorts of scientific equipment that should be sent off into space. Luisi (1998) discussed a variety of definitions of life and pointed out the definition adopted by the National Aeronautics and Space Administration (NASA) as a general working definition: 'Life is a self-sustained chemical system capable of undergoing Darwinian evolution' (reference citations in Luisi, 1998); he also indicated the limitations of this, particularly that proof of a Darwinian evolution may need thousands of years of observation and:

> . . . the notion of Darwinian evolution can only be applied to a population – therefore immediately excluding all single specimens, artifacts, chemical, and artificial life forms. Suppose that some NASA astronauts encountered single plants or single dangerous animals, they would not define them as alive, since there is no Darwinian population, and they may be eaten up in the meantime. Victims of a wrong definition of life! (Luisi, 2006, chapter 2)

This is also an aspect discussed at length by Hazen (2005, p. 27), who mentions the need for the likes of NASA to speculate on the full range of phenomena that might be said to be alive – robotic life, computer life, even a self-aware Internet – because a clear definition of life is essential for planning future missions in NASA's efforts to look for life on other worlds. This is not my concern here; instead I want to limit attention to planet Earth.

Let's see how far we can get by looking at current organisms to remind ourselves about common features that might be ancestral. We must start the argument by assuming that the property of 'being alive' is nothing more special than highly integrated chemistry. There is no value in ascribing some sort of 'vital spark' to biology that is beyond the grasp of established chemistry or basic physics because there's no point starting a discussion you expect to get out of your reach. All life that we know depends on a carbon-based chemistry and requires water as a universal solvent, so these must be considered primeval requirements on Earth, even though in other places and other times it is possible to imagine life perhaps based on silicon chemistry and/or using another polar solvent (such as liquid ammonia). If 'being alive' doesn't require a 'vital principle', then it follows that there is nothing special about the chemistry of carbon compounds other than the fact that they are constructed from carbon atoms. So the carbon compounds that are commonly found in living organisms, which are a relatively small subset of the immense catalogue of potential carbon compounds, are those that best carry out their function at their step in the highly integrated chemistry that is life.

This implies that a chemical evolution preceded the biological evolution. That chemical compounds compete for chemical reactions and those that are most easily made are so frequently encountered that they were used most readily, and/or those that worked best were used most readily. Either way there is a selection process and, when the reaction proceeds, there is a measure of relative success in that selection – an equivalent of the Darwinian aphorism 'survival of the fittest'. For example, there are hundreds of amino acids because any organic compound with both a carboxylic acid group (–COOH) and an amino group (–NH_2) is, by definition, an amino acid, but out of this great diversity living organisms use only 20 different kinds of amino acids in their proteins (and another half dozen or so in their general metabolism). It's highly probable that many other amino acids have been tried in various roles at some stage during the origin of life 3.8 billion years ago but were more difficult to synthesise, less stable or otherwise less suitable than the roughly two dozen or so in use today. The same must apply to the relatively limited number of other rather simple molecules (sugars, fatty acids, purines, pyrimidines) that are used in polymers (polysaccharides, fats and oils, RNA and DNA) today. The essentials of present-day biochemistry were probably established very early during the origin of life and probably used the organic molecules that dominated the prebiotic world. Those molecules would have been the ones that were most readily synthesised autocatalytically.

Similar arguments lead to the conclusion that temperatures and pressures that allow life are unlikely to deviate greatly from those of the Earth's current climate. Present-day living organisms are limited to a narrow blend of temperature and pressure. At the Earth's surface where the atmospheric pressure is characteristically about one bar (the unit 'bar' is defined in the legend to Fig. 3.2) this means from slightly below 0 °C and up to 100 °C for prokaryotes and about 0 to 55 °C for eukaryotes. In 'deep–hot' environments, extreme thermophile prokaryotes thrive at higher pressures, and temperatures up to 115 °C, such as in cracks in rocks about one kilometre beneath the surface, or around hydrothermal vents in the deep ocean. But 115 °C appears to be an absolute limit that requires evolution of special enzymes with enhanced temperature stability and membrane lipids (branched lipids) with higher melting points. Despite the indications of some molecular phylogenies that thermophiles may be ancient (and the consequent designation of these thermophiles as 'archaea') this need for evolutionary adaptation across a wide range of dissimilar enzymes together with a coordinated evolutionary adaptation in membrane structure is the main reason I think it unlikely that such organisms truly represent an ancestral lifestyle. The more likely scenario is that the earliest stages in the origin of life took place in relatively benign conditions: probably within the temperature range 0 to 100 °C and at a local pressure of about one bar.

There are two essentially physiological properties of current living organisms that are probably ancestral:

(a) Life is cellular, with a distinct population of 'inside' molecules separated from the general 'outside' by a cell membrane. This membrane is a lipid bilayer; constructed so that the hydrophobic tail regions of the lipid monomers are grouped in the inner part of the membrane with their hydrophilic head regions, which are 'polar' because they bear electrochemical charges, exposed to the aqueous environment at the outer layer and the aqueous cell contents at the inner layer. Any ions or molecules with an electrochemical charge cannot diffuse freely through the central hydrophobic zone of the membrane bilayer. This is what enables the membrane to be selectively permeable to ions and organic molecules and control the movement of these into and out of the cell. Controlled selective permeability, together with regulation of the internal chemistry, contributes to the maintenance within narrow limits of the interior chemical environment of the cell, even when the external environment is variable. A cell that can no longer regulate its internal environment will die.

(b) The universal energy metabolism of present-day cells depends on a pump that transports protons (H^+) across a membrane, thereby establishing an electrochemical gradient (more protons on one side than on the other). This creates a type of potential energy available for chemical work because the compensating return flux of H^+ can be coupled to the selective unidirectional transport of metabolites across the membrane and/or to synthesis of adenosine triphosphate (ATP), the molecule that is used as the universal energy-carrier in the metabolism of all present-day organisms. This electrochemical potential difference between the two sides of a membrane has been variously called an ion gradient, chemiosmotic potential, or proton motive force (see Fig. 9.2). It is today common to prokaryotes and eukaryotes alike, and in both types of organism is likely to be adapted locally to specific purposes in the cell membrane or the membranes of mitochondria, chloroplasts, and other membranous compartments. Other ions (particularly Na^+ and K^+) may also be pumped to establish an electrochemical gradient. However, the present-day universality of the proton gradient, and its specific coupling to ATP synthesis, imply that both these features were ancestral in the origin of life (Lane, Allen & Martin, 2010).

Some aspects of modern cell metabolism are also so universally used that they may indicate ancestral events. Of course, in modern cells all aspects of metabolism depend on enzymes for their progress at rates required for the maintenance of life. However, some of the common metabolic cycles, such as parts of the tricarboxylic acid (or Krebs) cycle, have been modelled as autocatalytic cycles in which the thermodynamic relationships between the reactants favour a 'self-propelled' (that's what autocatalytic means) cycle of reactions of the sort that might have occurred before the origin of living things. This is not to say that autocatalytic cycles would have been particularly efficient or speedy, but they could have been the starting point from which greater efficiency or improved rate evolved with time. The first enzymes could well have been inorganic chemicals and/or metal ions acting as catalysts and, as Gánti (2003, p. 63) pointed out, enzymes impose direction and order, and this may have been their main initial function, rather than increased reaction rate. I can envisage selection pressure in favour of a pathway being expressed non-biologically as a preferential direction and/or a limit on side-reactions; and this being more important in the initial stages than an amplified rate for a pathway that otherwise still had undiminished alternative directions or side-reactions. In the beginning, what happens is much more important than how fast it happens. The potential for autocatalytic cycles in the origin of life is further discussed in Chapter 9.

If an autocatalytic cycle works it persists and if it doesn't work it stops; and while it is stopped its component organic molecules may be used (= recycled) for something else so that the original cycle dissipates (= dies). There is, therefore, a chemical selection process which is the essence of chemical evolution. But this is not Darwinian evolution. And it is Darwinian evolution that must be applied to even the simplest of living organisms. Darwinian evolution is a natural universal law that applies to organisms with the following three properties:

(a) **Reproduction and heredity.** Organisms reproduce (not necessarily by a sexual process, it could be a simple division), but the properties of the parents are inherited by the offspring. The properties of individuals depend on some sort of heredity based on a genetic message which is conveyed from parent to offspring.
(b) **Variation and mutations.** Inherited variation can occur in the genetic message, which may result from its being imperfectly replicated and/or randomly changed (mutated). The result is variation in the properties of the offspring.
(c) **Differential fitness and selection.** Some of the inherited variants are better fitted to their environments; this might be expressed as faster growth rate, lower mortality, etc., that leads to greater reproductive success and a consequent increase in number relative to other varieties. Fitness is a function of the impact of the environment on the organism and is expressed as reproductive success.

In all present-day cells the genetic information is carried in DNA molecules, built from four nucleotides, which are phosphate-deoxyribose molecules attached to one of four heterocyclic (that is, the ring-form molecules contain both carbon and nitrogen) organic bases, namely the purines adenine or guanine, or the pyrimidines thymine or cytosine. In all present-day cells the DNA polymer is a duplex, meaning that natural molecules are actually composed of two DNA chains in a head-to-tail association with the nucleotide bases in the middle of the duplex and arranged in characteristic hydrogen-bonded pairs of 'complementary bases': adenine with thymine and guanine with cytosine. The result is that in all modern organisms the DNA duplex (the 'double helix') is made up of two complementary chains.

This provides a strategy for the precise replication of long DNA chains: separate the duplex into two separate molecules, build a complementary copy on each molecule using the standard complementary base-pairing rules, and the result is two daughter copies of the original duplex. In all present-day cells:

(a) the genetic information is carried in the form of the sequence of bases along the length of the DNA molecules in the duplex;
(b) the DNA code is a representation of the linear arrangement of amino acids in the proteins to which each gene corresponds;
(c) three nucleotide bases make up a code sequence for each amino acid in the protein (i.e. the 'codons' of the standard 'triplet code', thus there are a total of 64 codons because there are three positions in each codon, in each of which there can be one of four different organic bases [= 4^3 possible combinations]);
(d) the same coding dictionary is used (that is, which triplet corresponds to which amino acid is standardised) across all extant organisms.

The translation process that converts the genetic message into working protein is also uniform across present-day organisms. In all, it depends on transcription of one molecule of the DNA duplex (the 'sense' strand) into a complementary RNA strand, which is either immediately (in prokaryotes) or eventually becomes after processing (in eukaryotes) the messenger-RNA (mRNA) molecule. This mRNA is then conveyed to the ribosomes, complex molecular machines made up of many different RNA and protein molecules that are responsible for translating the linear genetic code of the mRNA into the linear amino acid sequence of the protein. They do this using the transfer-RNA (tRNA) adapter molecules, which have an anticodon complementary to one of the genetic code's codons at one end and the amino acid to which that codon corresponds attached to the other. Today there are approximately as many tRNAs as there are codons (that is, 64); though some cells have fewer because they have evolved to specialise on a restricted set of codons, and some have more because they have evolved to use different sets of tRNAs for different aspects of their differentiation. The amino acid is attached to its tRNA by one of 20 tRNA activating enzymes (officially, aminoacyl-tRNA synthetases) that recognise the anticodon and use ATP to bind the appropriate amino acid to the other end of the tRNA. Clearly, the proper working of the genetic code as it exists today is entirely dependent on the precision of the aminoacyl-tRNA synthetases; if they don't get it right when they activate their tRNA then the translation of the genetic code will be garbled.

It is assumed that the complete genetic machine (meaning the full set of DNA, RNA, translation apparatus, and transfer-RNA activation) that is common to all present-day organisms must be, because it is common to all, ancestral to all present-day organisms. It is not particularly easy to understand how the system evolved. Because what we see today is so thoroughly integrated, a 'primitive' version is difficult to envisage. Despite the best efforts of some of the most eminent theorists, the manner of the evolution of this 'genetic machine' remains one of the major unanswered questions about the origin of life. There have been some attempts worth noting.

I started this chapter with a quotation from Daniel E. Koshland's essay entitled 'The Seven Pillars of Life' (Koshland, 2002). He defined 'pillars' as the essential principles, thermodynamic and kinetic, by which a living system operates, and offered 'a living organism is an organized unit, which can carry out metabolic reactions, defend itself against injury, respond to stimuli, and has the capacity to be at least a partner in reproduction' as his definition, though said he was not happy with such a brief one. More extensive reflection resulted in these seven pillars:

(a) a Program, meaning an organised plan that describes the ingredients and the kinetics of interactions between ingredients; in the present day this program is implemented by the DNA;
(b) the second pillar is Improvisation, by which he means response to the environment, but which he specifically understands as mutation plus selection in the present day;
(c) the third of Koshland's pillars of life is Compartmentalization, which he infers from the fact that all organisms alive today have membrane-enclosed cells;
(d) the fourth pillar is Energy, because he claims that life as we know it involves movement (of chemicals, of the body, of components of the body) and must therefore be an open, metabolising system;
(e) Regeneration is his fifth pillar, because metabolic losses will change the system unless there is a compensating regeneration system, one of his examples being resynthesis of constituents subject to wear and tear, though he also includes cell division and reproduction under this heading;
(f) Koshland's sixth pillar is Adaptability, which he offers as a more immediate response to the environment than Improvisation, and he instances response to pain in humans and explicitly states that 'adaptability (pillar number 6) is a behavioral response';
(g) finally, the seventh pillar is Seclusion, by which he says he means 'something rather like privacy in the social world of our universe' although I interpret his meaning as 'specificity' because he gives as example 'enzymes that work only on the molecules for which they were designed and are not confused by collisions with miscellaneous molecules from other pathways ...'. (Koshland, 2002).

I have capitalised the names given to Koshland's seven pillars because the acronym which results is important to him: 'These seven pillars of life – P(rogram), I(mprovisation), C(ompartmentalization), E(nergy), R(egeneration), A(daptability), S(eclusion), PICERAS, for short – are the fundamental principles on which a living system is based.' He also states at the beginning of his essay 'If I were in ancient Greece, I would create a goddess of life whom I would call PICERAS, for reasons that will become clear.' A stated goal of this universal definition is to aid in understanding and identifying artificial and extraterrestrial life: 'Current interest in discovering life in other galaxies and in recreating life in artificial systems indicates that it would be desirable to elucidate those pillars'. I think Koshland's argument is completely animal-centred, indeed human-animal-centred, and for this reason do not find his seven pillars particularly useful, even for identifying artificial and extraterrestrial life.

Hazen (2005, pp. 26–31), in an extensive discussion, identifies the problem of the bias of the writer by drawing attention to the fact that to define the exact point at which a system of gradually increasing (= gradually evolving) complexity becomes 'alive' is intrinsically arbitrary:

> Where you, or I, or anyone else chooses to draw such a line is more a question of perceived value than of science. Do you value the intrinsic isolation of each living thing? Then for you, life's origin may correspond to the entrapment of chemicals by a flexible cell-like membrane. Or is reproduction – the extraordinary ability of one creature to become two and more – your thing. Then self-replication becomes the demarcation point. Many scientists value information as the key and argue that life began with a genetic mechanism that passed information from one generation to the next ... (Hazen, 2005, p. 29)

Gánti (2003, chapter 3, pp. 75–80) avoids the necessity of including, as so many authors do, a long list of properties in a definition of life by discussing the criteria that must be fulfilled for a system to be described as living. He starts from the premise that life is a function of material systems organised in a particular way. That is, life is the property of specially organised systems. The 'special organisation' is the possession of a characteristic set of processes and it is the operation of the whole system that results in the distinguishing events that we use to separate living from non-living in our everyday experience. Gánti (2003) makes the important distinction between a living system that is dead ('Death is an irreversible change which makes the system irreversibly incapable of functioning') and one that is quiescent, not functioning but capable of functioning if suitable circumstances arise ('This latter state corresponds to latent life, clinical death, resting seeds, dried-out micro-organisms, and frozen organisms. This is a state which is non-living but not dead'). Gánti divides his criteria into two classes: **absolute life criteria** and **potential life criteria.**

Absolute life criteria are essential attributes of individual living systems (each being an 'unavoidable criterion of the living state') and comprise:

(a) A living system must inherently be an individual unit, and is regarded a unit (a 'whole') if its properties cannot be additively composed from the properties of its parts, and if this unit cannot be subdivided into parts carrying the properties of the whole system. A system forming a unit (unit system) is not a simple union of its elements, but a new entity carrying new qualitative properties compared with the properties of its parts.
(b) A living system has to perform metabolism, 'metabolism' being the active or passive entrance of material and energy into the system which transforms them by chemical processes into its own internal constituents. Waste products are also produced, so that the chemical reactions result in a regulated and controlled increase of the inner constituents as well as in the energy supply of the system. The waste products eventually leave the system, either actively or passively.
(c) A living system must be inherently stable; this stability is not identical with either the equilibrium or the stationary state. It is a special organizational state of the system's internal processes, which makes the continuous functioning of the system possible and remains constant despite changes in the external environment. It means that the system as a whole, although continuously reacting via dynamic changes occurring within the living system, always remains the same. Further, it means that despite the permanent chemical transformations occurring in the living system, the system itself does not decompose; rather, it grows if necessary.
(d) A living system must have a subsystem carrying information which is useful for the whole system; this is the information necessary for the origin, development, and function of the system.
(e) Processes in living systems must be regulated and controlled; regulation in living systems occurs primarily through chemical mechanisms. Regulation may not be a separate criterion, since neither metabolism nor stability can be realised without regulation of the system's processes; regulation is already implied by these criteria.

Potential life criteria are applicable to the population rather than the individual, and 'whose presence is not a necessary criterion for individual organisms to be in the living state, but which are indispensable to the survival of the living world':

(a) A living system must be capable of growth and multiplication.
(b) A living system must have the capacity for hereditary change and for evolution, which is the property of producing increasingly complex and differentiated forms over a very long series of successive generations as the result of Darwinian selection.

(c) Living systems must be mortal; death is indispensable to life both to ensure recycling of organic material and to ensure the expression of selective advantage in successive generations. (Modified from Gánti, 2003, chapter 3, pp. 76–80.)

Luisi (1998) anticipates Hazen's comment (2005; quoted above) about 'arguments and ... hard-line positions' and discusses 'how the different definitions of life reflect the main schools of thought that presently dominate the field on the origin of life.' In other words, when you read a published definition of life you should seek out the author's prejudices and the background context from which the definition emerges. For example, Gánti (2003) states:

> At least two different kinds of life exist. I would like to repeat and emphasize this: we can identify at least two different kinds of life. I must emphasize that this statement is not a hypothesis but the everyday experience of biology and medicine. However, to the best of my knowledge, nobody has yet recognized, or at least stated this fact ... We know that after the death of an animal its organs, tissues, and cells remain alive for a time ... If we consider death as nothing other than the irreversible end of a life, then logically we must conclude that the life of a human or an animal and the life of its organs, tissues or cells are not the same ...

and

> Each living animal represents two kinds of life at each moment of its existence: the primary life of its cells and the secondary life of its body. The secondary life is the real life of the animal ... (Gánti, 2003, chapter 1)

These quotations reveal Gánti's fixation on animals, to which I have referred before, but relatively little additional thought is required to realise that the descriptive statement is applicable to all multicellular eukaryotes; animals, plants and fungi. In the rest of this book I will rely on the life criteria of Tibor Gánti (2003) to track the emergence of life, with suitable adjustment in the light of the discussions published by Luisi (1998) and Hazen (2005).

I ended Chapter 6 with a quotation from Robert Hazen's book *Genesis: The Scientific Quest for Life's Origin*, and I will end this chapter with another such quotation because it sets the scene so well:

> Attempts to formulate an absolute definition that distinguishes between life and nonlife represents a ... false dichotomy. Here's why. The first cell did not just appear, fully formed with all its chemical sophistication and genetic machinery. Rather, life must have arisen through a sequence of emergent events – diverse processes of organic synthesis, followed by molecular selection, concentration, encapsulation, and organization into diverse molecular structures. The emergence of self-replicating molecules

> of increasing complexity and mutability led to molecular evolution through the process of natural selection, driven by competition for limited raw materials. That sequential process is an organizing theme of this book.
>
> What appears to us as a yawning divide between nonlife and life obscures the fact that the chemical evolution of life occurred in this stepwise sequence of successively more complex stages of emergence. When modern cells emerged, they quickly consumed virtually all traces of the earlier stages of chemical evolution. 'Protolife' became a rich source of food, wiped clean by the consuming cellular life, like a clever murderer leaving the scene of the crime ... (Hazen, 2005, chapter 2, pp. 28–29)

But bringing something to life is not a crime. It's not the scene of a crime we seek but the scene of a great victory in a life and death contest of probabilities in which all the players compete with each other and against their environment to show who is the fittest; who is most favoured to survive. In a sense it's the life game. We are beginning to identify the players and in the next chapter will assemble the teams. After that we need to locate the field of play where the first part of the game is played out; where the die is cast; and where we find out the probability that life can win.

NINE

COMING ALIVE: WHAT HAPPENED AND WHERE?

In the narrative of life on Earth we have now reached the Archaean Eon (3.8 to 2.5 billion years ago); the age when chemistry came alive. During this period the Earth day increased from about 15 hours long to about 18 hours and the Sun brightened to 80% of its current level. At present, few data are available that are able to specify the atmospheric, oceanic or geological conditions on the early (prebiological) Earth. It can be reasonably assumed that conditions were very hostile due to volcanism, radiation, and continued bombardment by objects large and small from space; but it is likely that the average climate was temperate rather than extremely cold or hot (Kasting & Howard, 2006). As I have discussed in Chapter 6, there is no agreement on the gaseous composition of the primeval atmosphere apart from the general acceptance that oxygen was absent. According to Lazcano & Miller (1996): 'atmospheric chemists mostly favor high CO_2 + N_2, whereas prebiotic chemists mostly favor more reducing conditions'. In fact high levels of carbon dioxide in the early atmosphere are indicated by the high level of carbonate minerals in rocks of that age.

One of the reasons for postulating reduced gases in the atmosphere of the early Earth is that such an atmosphere would have had a greenhouse effect, which would compensate for the dimmer Sun of the day and keep the Earth from freezing over (Sagan & Chyba, 1997). It is thought that there was a global glaciation in the mid Archaean, approximately 2.9 billion years ago, which might have been caused by a hydrocarbon haze shielding solar radiation, the haze resulting from methane photolysis after the newly evolved methanogens increased the atmospheric methane/carbon dioxide ratio. This glaciation ended as continued production of methane produced greenhouse warming. Subsequent to the success of photosynthetic organisms, increase in atmospheric oxygen decreased greenhouse warming by methane and probably caused global glaciation just after the end of the Archaean (approximately 2.4 billion years ago) in the early Palaeoproterozoic (Kasting & Howard, 2006). These glaciations show the fine balance required to avoid global freezing ('snowball Earth'; see Fig. 11.2) primarily because of the faint early Sun. Sagan & Chyba (1997) calculated that an atmospheric mixing ratio of ammonia (the mixing ratio C_X of a gas X is defined as the number of moles of X per mole of air) of only about 10^{-6} to 10^{-4} would have been sufficient on the early Earth to cause enough greenhouse warming to counteract the effects of reduced solar radiation.

The primitive atmosphere most probably consisted mainly of nitrogen (75%) and carbon dioxide (15%), with minor amounts of various oxides of nitrogen and sulphur, methane and probably ammonia (Nisbet, 2000). Kasting & Howard (2006) concluded that the early Earth was warm, rather than hot, and that general conditions were temperate, though they point out that a moderate climate does not preclude the existence of both low-temperature (perhaps high-altitude?) and high-temperature (volcanic) environments across the Earth. Similarly, Lazcano & Miller (1996) state that processes relevant to the origin of life may have taken place in environments different from the terrestrial average, and instance hot springs, eutectic seawater, and drying lagoons as examples where the chemical conditions could be very different from the average. Nisbet (2000) outlined the environments in which life thrived in the late Archaean in a sketch map of Archaean ecology, indicating places where early life may have flourished as:

(a) shallow waters (oxygenic and anoxygenic photosynthesis);
(b) muds in the sea floor (potential for evolution of methanogens and exploitation of the sulphur cycle);
(c) alkaline vent shallow-water hydrothermal systems (potential for mixed chemotrophism and photosynthesis);
(d) mid-ocean ridges and deep-water hydrothermal systems (potential for chemotrophic communities based on hyperthermophiles);
(e) the open ocean (photosynthetic plankton).

There is reason to believe that early bacteria were able to thrive in all these different environments, so the list demonstrates that the diverse habitats are not alternatives because they are contemporaneous, and might imply that prebiotic activities also most likely occurred in several locations and environments at the same time (Bernstein, 2006; Nisbet, 2000). It is important to stress this point because most writers on the topic of the origin of life have some particular line of argument to promote and many of them present their own line of argument as 'the one true path'.

In my view, if a coherent case can be made for some steps of prebiotic chemistry occurring in a black smoker, then they probably did occur. Similarly, if a case can be made for prebiotic synthesis of amino acids, purines and pyrimidines as a result of spark discharges in reducing gas mixtures, then they also probably occurred (and if the average atmosphere was not sufficiently reducing then there is always the heavy lightning discharges that occur in the gas plumes of volcanoes, which, if they were anything like today, could have contained hydrogen, hydrogen sulphide, methane and carbon monoxide). There's a good case, too, for organic synthesis and the formation of chemiosmotic vesicles around shallow water and terrestrial alkaline hydrothermal vents. And what contribution is made by all that extraterrestrial material raining down from outer space? Significant amounts of biogenic compounds could be delivered by a heavy shower of interplanetary dust particles drifting through the atmosphere to land in a tiny pool of rainwater on the slopes of a volcano. Even a particularly well-endowed meteorite doesn't have to leave a smoking crater. Some large pieces of the Murchison meteorite 'were found on a road and the largest one weighing 680 g came through a roof and fell in the hay' (for further details view the following URL: www.lpi.usra.edu/meteor/metbull.php?code=16875). Imagine such a meteorite clattering down the slope of a volcano rather than through a barn roof, and coming to rest in a small pool of water. The Murchison meteorite contained water-soluble amino acids, purines, pyrimidines, and fatty acids that self-assemble into bilayered vesicles (Deamer, 1985; Deamer *et al.*, 2002; Deamer & Weber, 2010); just about everything we need.

As well as robustly debating the location for the origin of life, there is a major division between those who postulate that the first 'organisms' were heterotrophic and those who argue they were autotrophic. Darwin's warm little pond and Haldane's dilute soup of an ocean suggested the rather easier route to a heterotrophic organism. The Miller–Urey experiments revealed a constant source of nutrients for the evolving heterotrophs and this notion was widely accepted for many years. But ideas about the atmosphere of the primitive Earth changed radically and the carbon-dioxide-rich model postulated such a dense carbon dioxide atmosphere (10 to 100 times greater than present-day atmospheric pressure) that it was virtually impossible for many people to envisage the spark-discharge synthesis of organic compounds in sufficient quantity to support heterotrophs (Kasting, 1993).

Autotrophic theories must involve non-enzymatic biosynthetic reactions between organic molecules emulating the pathways of intermediary metabolism of present-day organisms. Several schemes have been proposed. One of the first such proposals was that the citric acid cycle started with acetyl coenzyme-A (acetyl-CoA) by two carbon dioxide fixations in a scheme which:

> ... requires clays, transition state metals, disulphide and dithiols, U.V. and cyanide ion. A general scheme is proposed, involving the fixation of CO_2 and N_2, for the evolution of intermediary metabolism based on the evolution of a complex system from a simple one. The basic conclusion is that metabolism could have evolved from a simple environment rather than from a complex one. (Hartman, 1975)

Clays (and other minerals) are commonly involved in schemes for prebiotic 'metabolism' to serve as catalysts; similarly there is a need for external sources of energy (UV in this instance; spark discharges, and temperature and ion gradients in others), and reducing agents. I have already mentioned some of the more recent schemes. The most intricate autotrophic scheme is that of Wächtershäuser (1988, 1992, 2000, 2006), which puts the origin of metabolism in and around deep, hot hydrothermal systems (black smokers). In this scheme biosynthesis and polymerisation are postulated to take place on the surface of FeS and FeS_2. The reaction $FeS + H_2S \rightarrow FeS_2 + H_2$ is thermodynamically favourable, so the FeS/H_2S combination is a strong reducing agent. It is argued that in the pressurised iron-sulphur world of a black smoker NiS and FeS are coprecipitated into an aqueous slurry, which will also contain small (catalytic) amounts of other metals like selenium, cobalt and molybdenum. In the elevated temperatures and pressures of the hydrothermal system (equivalent to 100 to 250 °C and 0.2 to 200 MPa [2 bar to 2000 bar] in the laboratory) volcanic gases are converted on the surfaces of these sulphide slurry particles into low molecular weight organic compounds that today serve as intermediates in cell metabolism (Fig. 9.1).

For example, carbon fixation occurs when carbon monoxide (CO) and hydrogen sulphide (H_2S) react on the NiS/FeS coprecipitates, forming methanethiol (CH_3SH) which then reacts with more CO to form the activated thiolester CH_3-CO-SCH_3, which hydrolyses to acetic acid (Huber & Wächtershäuser, 1997). Similarly, pyruvic acid (CH_3-CO-COOH) was synthesised by reaction of thiols and carbon monoxide (formic acid being used as source of CO) with iron sulphide (Cody *et al.*, 2000). Thiolesters, such as acetyl coenzyme-A, play crucially important roles in present-day biological metabolism where they are employed as acyl transfer reagents for the synthesis of other metabolites so their easy synthesis in the iron-sulphur world is extremely important.

And the potential natural synthesis of pyruvic acid where reduced hydrothermal fluids pass through an iron sulphide crust could have provided the prebiotic Earth with critical primordial initiation reactions for a chemoautotrophic origin of life (Wächtershäuser, 2000). In this paper Wächtershäuser stresses that pyruvate is too unstable ever to be considered as a slowly accumulating component of a prebiotic soup, but is readily

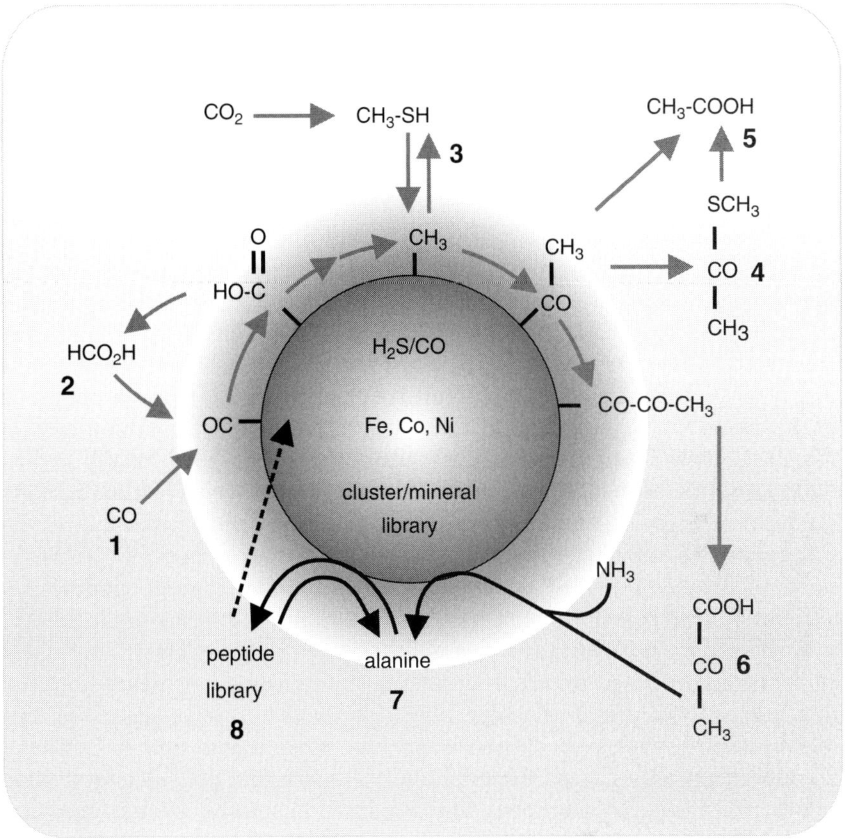

Fig. 9.1 Reactions in the iron-sulphur world. The illustration shows how carbon fixation occurs when carbon monoxide (CO) and H_2S react on the surface of the mineral NiS/FeS coprecipitate, which includes other metals and is described here as a 'mineral library'. All individual reaction steps converting carbon monoxide [1] to peptides [8], including formation of methyl thioacetate [4], pyruvate [6], of alanine [7] by reductive amination of pyruvate, and of peptides [8] by activation of amino acids with CO/H_2S have now been demonstrated in the laboratory under high temperatures and pressures similar to those found in deep-ocean hydrothermal vent systems (equivalent to 100 to 250 °C, and 0.2 to 200 MPa [2 bar to 2000 bar] in the laboratory). Autocatalysis and prebiotic evolution may involve a primitive version of the citric acid (Krebs) cycle in which (methyl) thioacetate and pyruvate participate and/or peptide feed back to the catalytic metal centre (dotted arrow). (Redrawn and adapted from Wächtershäuser, 2000.)

formed in the surface metabolism taking place in the environment of positively charged iron sulphide mineral surface at hot hydrothermal vents. He emphasises that the two-dimensional order of these hot pressurised mineral systems suits them to the creation of long serial reaction cascades and catalytic feedback loops from the start of metabolism and produces a maxim 'order out of order out of order' for the iron-sulphur world to contrast with an 'order out of chaos' maxim for the prebiotic soup approach.

So far, experimental results at high pressure and high temperature support the view that primordial organisms could have been autotrophs feeding on carbon monoxide. But the reactions shown in Fig. 9.1 would also be a source of geo-organic compounds that may serve as food for heterotrophs. On early Earth, outgassing (the release of gases by volcanic activity) occurred on a massive scale and over an enormous extent of the surface, becoming restricted to vents and volcanoes as the crust thickened. Indeed, the conditions are still available on Earth today, though they are less frequently encountered than they were in the past; they may thus be a source of geo-organic compounds that may serve as nutrients for present-day heterotrophs. There are criticisms of the overall concept, though. Lazcano & Miller (1996) claim that carbon dioxide is not reduced to amino acids, purines, or pyrimidines in deep hydrothermal systems, so: 'it seems unlikely that the vents played a role in prebiotic synthesis of organic compounds or polymers'. The scheme also requires the first organisms to be hyperthermophiles but there is no agreement on this. As Lazcano & Miller (1996) put it: 'Hyperthermophiles may be cladistically ancient, but they are hardly primitive relative to the first living organisms'.

What is to me the most satisfying autotrophic scheme is that which Russell *et al.* (1994) proposed for the hydrothermal vents known as 'white smokers' that release cool, alkaline water rich in hydrogen. I have described this scheme above, so here will only make the point that I find it satisfying because it not only has all the sulphidic chemistry that provides for a wide catalogue of organic syntheses, but it features a natural proton gradient, across a mineral membrane around iron sulphide bubbles (Martin & Russell, 2007; Russell & Hall, 2002). The proton gradient is the universal chemiosmotic energiser in all organisms of the present day, despite the fact that other ions might have been used. Being universal implies being primeval and that's why I find the natural formation of such a gradient in bubbles released from alkaline vents so satisfying (Fig. 9.2).

However, although chemiosmotic coupling may have evolved around alkaline vents I am not convinced that autotrophic schemes provide the best explanation for the first step towards life. These specific autotrophic schemes see life originating in the ocean and that's something with which I can't agree because, as I have already indicated, the origin of life, however momentous, was a physically tiny event. Proponents of autotrophism argue that the difficulty of accumulating organic compounds of abiotic origin in the primitive ocean to concentrations able to make it possible for life to arise is a severe barrier to a heterotrophic origin of life. That barrier, in their view, then becomes insurmountable if the general atmosphere is not a chemically reducing one. As I have already indicated above, these criticisms are negated

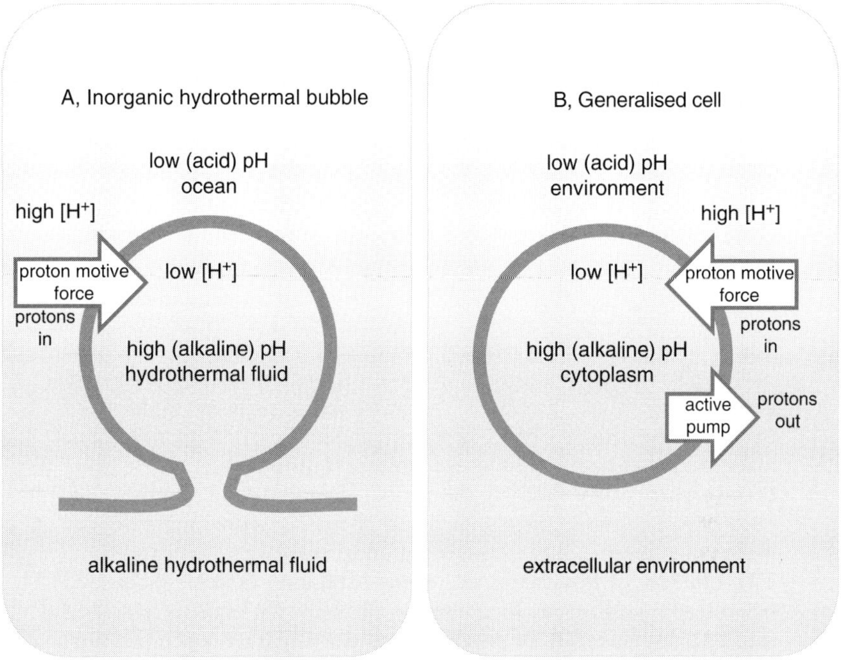

Fig. 9.2 Chemiosmotic properties of cells. **A**, the proton-motive force across the boundary of a hydrothermal bubble with a mineral membrane. The proton-motive force is a gradient of H^+ concentration and electrical potential that stores energy and makes it available for synthesis and transport. The proton-motive force is made by an alkaline (high pH) internal effluent from the bubble's hydrothermal vent jetting its fluid into the acidic (low pH) external ocean environment, which is essentially a carbonic acid solution. **B**, the proton-motive force of generalised living cells descended from LUCA. In this case the proton-motive force is made by an alkaline (high pH) cytoplasm and by an acidic (low pH) extracellular environment. The cell itself creates and maintains the proton gradient by active (that is, energy-dependent) pumping of protons across the cell membrane and out of the cell. (Redrawn and modified from Lane, Allen & Martin, 2010.)

by the assumptions (a) that the chemical reactions leading towards the living state take place in aerosol droplets, and (b) that Miller–Urey-type synthesis takes place in the gas plumes and dust clouds emerging from volcanoes. If we add the enormous quantities of organic compounds arriving on Earth from space, then the heterotrophic route for the origin of life could be restored to the favoured position it held when Darwin, Oparin and Haldane first advocated the idea.

A prebiotic soup (even if contained in microlitre quantities in aerosols) contains most, if not all, of the necessary small molecules, and represents a large potential energy supply on the early Earth available to a number of different primitive metabolic pathways. 'It is clear that such compounds could provide both the growth and energy supply of a large number of organisms, but this would rapidly result in the depletion of the available nutrients' (Lazcano & Miller, 1996). As nutrients of prebiotic origin were depleted, the exclusively heterotrophic pre-alive systems would experience a metabolic crisis that would result in extreme selective advantage for the evolutionary development of light-harvesting autotrophic systems with carbon-dioxide-fixing abilities. Yuri Sorokin described this process elegantly in 1957:

> The driving force of evolutionary development from heterotrophy to autotrophy was the competition for organic nutrient substrate, the tendency of individual groups of heterotrophic organisms to be freed from the necessity of obtaining energy and carbon only from organic matter. The competition for the organic substrate appears to have become particularly acute at the period of evolution of heterotrophic microorganisms when these mineralized all the stores of primary organic matter of abiogenic origin, converting them to H_2, CH_4, H_2S and NH_3. The influence of this factor led to the adaptation of some pigmented heterotrophic organisms to utilization of their light-sensitive pigments for imparting by means of the energy of light a high reduction potential to the hydrogen of water and of some other compounds; and, in this way, to produce reducing agents of the type of $DPNH^+$ [now called reduced nicotinamide adenine dinucleotide: $NADH^+$]. These reductants participated in the reduction of CO_2 and their oxidation provided energy. In this way there appeared the photoautotrophs, i.e. photosynthesizing bacteria ... (Sorokin, 1957, p. 371)

Another important point is that the initial steps in this heterotrophic scheme do not need continuity. It is only when compartments form to contain complex metabolism that continuity becomes important, and by that time the pre-alive systems are well on the way towards being living cells. Before that, if, say, some amino acid is formed and then the droplet dries up; it doesn't matter. Again, by definition, there are no microbes around to degrade the chemical. It will remain, perhaps still in suspension in the atmosphere, perhaps as an organic smudge on some rocky surface, until some water vapour condenses and it dissolves again, possibly at a different concentration than before and perhaps in combination with a neighbouring smudge with which it can chemically react. And we have enough time as the individual reactions to synthesise prebiotic compounds have short half-lives compared with the several million years that are available (Lazcano & Miller, 1996).

All origin-of-life schemes anticipate that the interacting molecular components are restricted to some sort of compartment within the bounds of which some degree of order can be established from the disordered surrounding environment (for detailed discussions see Davies, 2006; Dyson, 1999; Fenchel, 2002; Gánti, 2003; Hazen, 2005; Luisi, 2006; Lurquin, 2003). This is usually a membrane-bound compartment, but mineral and crystal surfaces have also been postulated. Cairns-Smith (1982) proposed that microscopic crystals of the minerals contained in common clay might have been able to organise early metabolism. Clay microcrystals are usually flat plates with their two plane surfaces exposed to the aqueous environment. The crystals have a regular silicate lattice, but with an irregular distribution of embedded atoms of metals such as magnesium and aluminium. Metal ions embedded in such a crystal create patterns of ionic charges that might adsorb specific organic ('nutrient') molecules and catalyse specific chemical reactions. Further, it is conceivable that the crystal might direct new rounds of crystallisation that would replicate the information contained in its ionic patterns. So the crystal serves as both a catalyst of metabolism and store of genetic information. The primary reason for developing this model was to compensate for the perceived 'thin ocean soup'; selected molecules from the allegedly dilute solution of organic compounds could be concentrated by adsorption on the clay mineral surfaces and thereby brought into favourable contact with other potential reactants. At some stage in its evolution the mineral system converts to the present biological one using proteins, RNA and DNA by an unspecified process (called genetic takeover). No experimental support for these ideas has yet emerged.

Another mineral suggested as being the original site of organic synthesis is iron pyrite (FeS_2); this is the 'iron-sulphur world' hypothesis of Günter Wächtershäuser to which I have referred above (Wächtershäuser, 1988, 1992, 2006). In contrast, Russell's interpretation of the iron-sulphur world is that a similar chemical environment produces chemiosmotic vesicles bounded by a mineral membrane across which a proton gradient is produced (Martin & Russell, 2007; discussed above). The first membrane-bound compartments to be proposed were the coacervates suggested by Aleksandr Oparin as early as 1922 (see http://en.wikipedia.org/wiki/Alexander_Oparin) though this was first translated into English in 1938 (Oparin, 1957a). Oparin argued that life originated through a chemical evolution of organic compounds synthesised from a primeval (reducing) atmosphere of methane, ammonia, hydrogen and water vapour. Oparin conceived a way in which microscopic droplets form spontaneously from particular dilute organic solutions; he called these coacervates (from the Latin *coacervare*, meaning 'to assemble together or cluster'):

> ... the type of organization peculiar to life could originate only as a result of the evolution of a multimolecular organic system, separated

> from its environment by a distinct boundary, but constantly interacting with this environment in the manner of 'open' systems. Since, as evidenced by a number of features present-day protoplasm possesses a coacervate structure, the mentioned systems, which represent the starting point for the evolution leading to the origin of life, could have been coacervate drops. (Oparin, 1957b, p. 221)

In the context of the biochemical knowledge of the era in which the hypothesis was proposed, 1920s to 1950s, there were several candidate organic molecules from which coacervates might form, but coacervate later came to mean the droplets (lipid vesicles or liposomes) formed spontaneously in solutions of fatty acids, lipid and phospholipid molecules (Deamer, 1985; Deamer *et al.*, 2002; Deamer & Weber, 2010). Oparin suggested that different types of coacervates formed, containing different combinations of organic molecules from the ocean, and were then subjected to a selection process from which eventually a living system arose.

> Even the coacervate drops which were first formed in the waters of the primaeval ocean must have been able to incorporate in themselves the organic substances of the surrounding medium. All their subsequent evolution was based on the natural selection of those systems which could assimilate these substances most quickly and efficiently. (Oparin, 1957a, chapter IX, p. 400)

What might be happening within the lipid vesicles was suggested by Tibor Gánti's chemoton hypothesis (Gánti, 2003). A chemoton is a self-reproducing cycle of autocatalytic chemical reactions, operating in the interior of a membrane-bound vesicle (Fig. 9.3). In Tibor Gánti's own words:

> Let us assume that we have a molecule containing two carbon atoms (and several other atoms) which can react with a molecule containing one carbon atom. The product is a molecule with three carbon atoms. In a subsequent reaction this reacts again with a molecule containing one carbon atom, and thus we obtain a molecule with four carbon atoms. If this reacts again with a molecule with one carbon atom, a molecule with five carbon atoms is produced. Let us assume that this molecule is unstable and, as a final reaction step, it can split into a molecule with two carbon atoms and a molecule with three carbon atoms. If the molecule with the two carbon atoms is identical with the original molecule, then we have arrived at our starting point, i.e. we have obtained a chemical cycle since the end-product can pass through the reaction chain described again and again, by producing a molecule with three carbon atoms in each cycle. Thus we can say that the cycle is a chemical machine capable of the continuous production of molecules with three carbon atoms from molecules with one carbon atom.
>
> But what happens if the molecule with four carbon atoms in this reaction series is unstable and splits into two molecules with two carbon

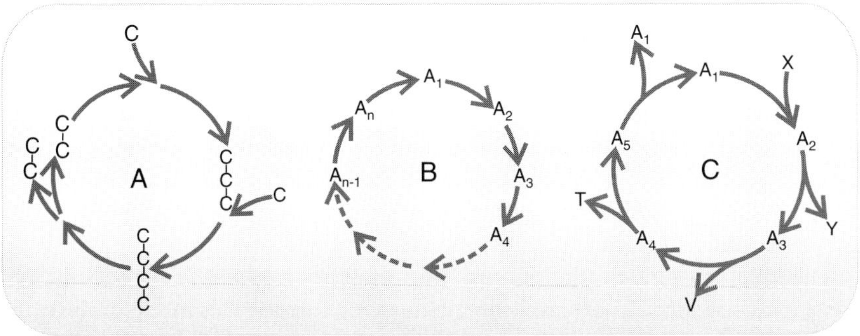

Fig. 9.3 Chemical cycles and chemotons. **A**, a simple outline chemical cycle (or self-reproducing fluid machine) involving an organic compound containing two carbon atoms (for example, glycolaldehyde). For simplicity only the carbon skeletons of the molecules are shown. This reacts with a molecule containing one carbon atom (formaldehyde) to form a molecule containing three carbon atoms (glycerolaldehyde) from which, in a further reaction, a compound with four carbon atoms (tetrose) is obtained. This molecule splits into two molecules each containing two carbon atoms (glycolaldehyde), thus reproducing the starting molecule. The process can continue with both product molecules, resulting in the production of 4, 8, 16, 32, 64 ... etc. molecules of glycolaldehyde in subsequent cycles. **B**, in a generalised chemical cycle the molecules are transformed in every step but the original molecules are restored at the end of the cycle. **C**, a self-reproducing minimal cycle for the chemoton. In addition to its own constituents (A_1, A_2, etc.) and waste materials (Y), the cycle also produces the raw materials for the informational subsystem (V) and membrane formation (T). If this system operates in the interior of a membrane-bound vesicle containing the information-carrying macromolecule, the molecules represented by V here are polymerised on the template present, and molecules T are inserted into the membrane. (Redrawn and modified from Gánti, 2003.)

> atoms (identical with the starting molecule) before it can react with a new molecule with one carbon atom? In this case we have a machine which, on the one hand, produces molecules with two carbon atoms, and, on the other hand, transforms itself back into a molecule with two carbon atoms. This means that at the end of the cycle we have produced two of our starting molecules. Now, both can react further; after another cycle there will be four molecules with two carbon atoms, after three cycles there will be eight, then 16, 32, 64, 128, etc., i.e. this molecule, which can go through a cyclic reaction, proliferates. (Gánti, 2003, chapter 2, p. 26)

Eventually, the different chemotons within a vesicle serve each other by producing substrates not available externally, or by transforming waste from one chemoton into substrate for another. In this way, by a process of evolutionary accretion an autocatalytic metabolic network is built up; the more complex it becomes, the more fixed it becomes. If the vesicle also contains an information-carrying macromolecule, and the means to insert molecules into the vesicle membrane then we have some of the requirements for a living, reproducing entity. Overall the integrated chemoton cycle makes use of nutrients and produces monomers for the information-carrying molecules, monomers for the membrane-forming molecules, waste materials and, usually in the final step as the chemoton's self-reproduction step, initial substrate molecules.

The heterotrophic origin just described assumes catabolic metabolism first, using external supplies of simple nutrients (which establishes the central core of the intermediary metabolism of present-day organisms) followed by selection in favour of an ever wider collection of catalysts (which might be exportable) that can digest (possibly externally) ever more complex nutrients in the surrounding environment. Vesicles are not formed singly, but in large numbers, possibly from the 10^{31} aerosol droplets formed during the period of origin (see Chapter 7), and we must build into this argument the notion that each vesicle (the aerosol droplets become vesicles when they resettle into a water interface; see Fig. 7.1) or aerosol droplet in a small volume of liquid will be in competition with every other vesicle. A Darwinian selection will be applied to this competition but positive selection may not initially be expressed in reproductive success, but rather in the persistence of the most successful vesicle(s).

Initially, vesicle reproduction will simply be a matter of a vesicle expanding ('growing') by insertion of membrane constituents into the membrane until the vesicle becomes physically unstable and separates into two. In such a case, components of the vesicle (that is, its chemotons and information-carrying molecules) would at first be distributed randomly between the 'daughter' vesicles, but there would be selective advantage in any system that ensured the two daughters were allocated the same (complete) collection of components as their successful parent. When this was established the Darwinian selection could be expressed in reproductive success. The catalysts employed in widening the nutrient supply by digesting large molecules might be simple chemical compounds, metal ion complexes, or even ribozymes (see below). The simplest and most easily available would be used first but the departure would ensure that selective advantage lay with ever increasing efficiency of the catalysts. Further, competition with other vesicles would favour catalysts that are bound to the membrane in some way, rather than being released freely into the environment. At the described level of development the chemoton vesicle has become a pre-alive system that satisfies some of the criteria for a living organism. So far I think the description of the pre-alive system given immediately above satisfies the first three of Gánti's absolute life criteria discussed earlier.

We now need to find a way to explain the evolution and integration of a subsystem carrying information necessary for the origin, development, and function of the system to satisfy Gánti's absolute life criterion-5. In modern-day organisms the living tasks are neatly assigned: DNA stores the information, RNA transfers the information to proteins, and proteins do the metabolic work. The severe problem is how these tightly integrated processes arose. It is relatively easy to understand how peptides, and eventually proteins, became involved in metabolism by being recruited as catalysts. Once recruited, competition between chemotons would result in selection pressure towards ever more efficient proteinaceous enzymes (the 'protein world'). Gánti (2003) comments about enzymes:

> In the living organism [of the present day], even the simplest prokaryotic cell, many chemical reactions run simultaneously. A given chemical substance is not restricted to a single chemical reaction; the various reactions are connected and form a complicated unified reaction network. For a biologist, this network is determined by the enzymes present in it. For a chemist, the basic properties of the network are determined by the transformations of the participating chemical compounds; the enzymes present in the cell merely determine the end result. (Gánti, 2003, chapter 3, p. 63)

This is an interesting and insightful interpretation that enzymes impose direction and order (and presumably also regulation and control) upon an otherwise chaotic network of potential chemical reactions. It has implications for selection pressure and evolution by its inference that enzymes could be acquired sequentially, and by viewing enzymes as directional determinants for an evolving metabolic pathway. To put some numerical sense on the size of the network we have to consider, Lazcano & Miller (1996) estimated the number of distinctive innovative enzymic reactions to be in the range 20 to 100:

> Most enzymes [in present-day organisms] are recognized to have arisen by gene duplication. The uncertainty is the number of enzymes that did not arise in this manner, i.e., the starter types. In some cases, the starter types may stem from slow non-enzymatic reactions where the protein improves on a previously sluggish process, e.g., pyridoxal catalyzed transaminations ... Based on the similarity of many biochemical reactions, and on the observation that many proteins of related function share the same ancestry within a given organism [in the present day], we estimate that the number of starter types ranged from 20–100, but the reader might want to make her or his own list of minimal enzymes. (Lazcano & Miller, 1996, p. 796)

The real problem lies in understanding how the nucleic acids became so intimately engaged in the process of metabolism.

> There are two schools of thought when it comes to the origin of life; 'genetics first' and 'metabolism first'. The 'genetics first' school is

> represented by Eigen & Schuster (1977; 1978a; 1978b) and by Maynard Smith & Szathamáry (1999), and the 'metabolism first' school by Oparin (1957a), Wächterhäuser, (2006), Cairns-Smith, (1982) and Dyson (1999). To a large extent this dichotomy is artificial and perhaps mainly a semantic problem, that is, a question on how life is defined … (Fenchel, 2002, p. 51)

Lazcano (2010) seems to agree with Fenchel's suggestion that the dichotomy may be artificial as he concludes that 'many of the observations and experimental findings that are used to argue in favor of one or another view are equally consistent with the premises of both theories and do not unambiguously support either of them.' While agreeing with this sentiment, I would numerically extend that 'either of them' to 'all of them' because this is more than a dichotomy of two origins; rather, this is the bringing together of many origins. This is something I want to stress as we move into the next chapter.

TEN

MY NAME IS LUCA

'The most striking feature of life is its similarity!' (Fenchel, 2002, chapter 12, p. 123). Cavalier-Smith (2010a) called it stasis and illustrated it this way:

> Explaining stasis is as important as explaining change ... Inheritance alone is too imperfect to achieve this. About half the nucleotides in ribosomal RNA (rRNA) molecules have an identical sequence in every bacterium, animal, plant and fungus, despite every nucleotide regularly mutating, some in every generation in every species. Since you started reading this paper, at least one cell of your body will have one or more new mutations in regions of rDNA where the ancestral sequence in the last common ancestor of all life has never actually been supplanted by evolution over 3.5 billion years. The same applies to hundreds of other genes essential for life. Stasis stems from the lethality (or dramatically lower fertility) of such variants (purifying selection) and is not inherent to the genetic material. Without death, life could not persist. Contrary to what Darwin thought, and many creationists still do, the problem is less to explain how genetic variation occurs, than to understand why some organismal properties never change while others frequently do. Differential reproductive success (anthropomorphically 'natural selection') biases genotypes of successive generations subjected to a perpetual, physically inevitable, barrage of mutations in every part of the genome. This beautifully explains both long-term stasis and radical organismal transformation. Both stasis and change are needed to explain the patterns of similarity and difference that enable hierarchical Linnean classification. (Cavalier-Smith, 2010a, p. 113)

The principle I have used in this book so far is the general consideration that features that are common to all organisms that exist today are the fundamental cell functions that today's organisms have inherited from their Last Universal Common Ancestor (LUCA).

These common features are the functional and structural properties of the cell membrane (energy transduction, transport proteins, etc.), the essential components of energy metabolism (particularly proton-gradient-driven chemiosmosis), a network of chemical interactions representing the core aspects of intermediary metabolism, and the genetic code and a genetic system based on DNA, the transcription of the genetic message into RNA, and the RNA-to-protein translation apparatus (see Chapter 8 for discussion). The ancestor may have had a cell wall or, perhaps, a looser arrangement of extracellular matrix molecules that could evolve on the one hand into a functional cell wall and on the other hand into a contribution to the biofilm matrix (see below). Together this set of properties map out the target that speculations about prebiotic events must aim for.

There is a difference of opinion about the type of energy metabolism employed by LUCA. Many favour the idea that a type of photosynthetic energy metabolism was used from the beginning, but I am uncomfortable with this. In Chapter 9 I have given a description of a heterotrophic origin in which the first chemotons used readily available nutrients and as these were exhausted chemical catalysts were developed for release into the environment to degrade the tars and other polymers accumulated by hundreds of millions of years of abiotic chemical reactions. This description prejudges any debate on which came first as this account clearly envisages metabolism arising first in the origin of life process. Metabolic chemotons are clearly expected to compete and be subjected to selection and, consequently, to evolve towards successively more intricately integrated metabolic systems that could even use membrane binding to partition chemoton cycles accurately between daughters when the membrane vesicles are inflated to a size that causes them to divide. The information in this system is embedded in the architecture of the system, so the challenge is to insinuate into this a nucleic acid-based subsystem that stores and interprets a different information language.

One way to introduce nucleic acids to this scheme is to suggest that some of the early enzymic activities were carried out by RNA molecules. Since Zaug & Cech (1986) described the discovery of sequence and pH-dependent RNA polymerase and ribonuclease enzymic activities in the RNA molecules of *Tetrahymena*, the RNA world has come to life (Atkins, Gesteland & Cech, 2010; Gilbert, 1986). This interpretation suggests that the first living organisms were self-replicating RNA molecules with catalytic activity (ribozymes) and phenotype and genotype both residing in the same polymer, and with no involvement of protein catalysts (Eigen & Schuster, 1977,

1978a, 1978b). Combining informational and catalytic properties into a single molecular species is an idea that has been widely accepted.

The route to this is envisaged to have involved a series of different genetic polymers; starting, perhaps, with the most likely prebiotic compounds such as ethylenediamine monoacetic acid, which could have provided the genetic polymer (called peptide nucleic acid, or PNA) with a backbone of *N*-(2-aminoethyl)-glycine units linked by peptide bonds instead of the sugar phosphates of today. Other analogues of nucleic acids have been laboratory synthesised, with hexoses rather than pentoses, or pyranoses replacing furanoses; these suggest that a variety of evolutionary experiments with informational macromolecules may have taken place (Robertson & Joyce, 2011). Indeed, riboswitches, which are widely distributed in modern bacteria and eukaryotes (Sudarsan, Barrick & Breaker, 2010), show that RNA molecules have a range of capabilities that suit them to metabolic roles rather than strictly genetic information transfer roles. Riboswitches are sequences that occur in the 5′-untranslated regions of messenger-RNAs that bind specifically to small molecules to regulate gene expression by affecting transcription or translation (Tucker & Breaker, 2005; Winkler *et al.*, 2004). Riboswitches selectively bind to metabolites like coenzymes, purines, pyrimidines, amino acids, and other small molecules and it has been suggested that they might be descended from ancient sensory and regulatory molecules (Breaker, 2011).

However, the fact that particular RNA sequences can bind specifically to simple metabolites prompts speculations about a more direct involvement of RNA in primeval metabolism. It is possible that some of the first hydrolase ribozymes could have been sent out of the pre-alive system's membrane vesicle to scavenge for 'nutrients' (such as sugars, dicarboxylic and amino acids) among the tars and other polymers left over from abiotic chemical synthesis. One could imagine the vesicle having a range of degradative ribozymes stored in inactive form (so they could not bind to essential internal polymers) by being parked by base-pair hydrogen bonding on an RNA polymer within the vesicle. It is also feasible that the ribozyme RNA was not fully released, but rather extruded through the membrane while anchored intracellularly by hydrogen bonding to its 'parking RNA' and then rewound by hydrogen bonding after it had picked up nutrient.

This overall function provides a rationale (an evolutionary logic) for ribozyme RNA evolving to greater specificity at both ends: greater specificity for nutrient improves the scavenging function and confers greater benefit on the pre-alive system; greater specificity between the two RNAs (i.e. emergence of codon–anticodon pairing) allows better placement of the ribozyme and also confers greater benefit on the vesicle. Lining up amino acid-binding ribozymes on a 'parking messenger-RNA' might also be an effective means of making a 'pseudo-peptide' before the emergence of mechanisms to covalently link the amino acids together. The mixed nucleotide/amino acid combination conferring some local feature (adhesion, hydrophobicity, ion selection, nutrient binding, etc.) on the membrane might confer selective advantage on the vesicle possessing it. With this basic principle embedded in the chemoton's metabolism, there would be selective value in evolving towards more and more efficient machinery that comes to resemble a specific 'messenger-RNA', combined with highly selective amino acid attachment to (what are now) transfer-RNA molecules, and (ribosomal) peptide/protein-assembly machinery. Ribonucleotides and amino acids may well have co-operated synergistically in these processes. Carny & Gazit (2005) point out that dipeptides can act as catalysts and self-assemble into regular tubular, fibrillar and closed-cage structures, and that tripeptides can act as templates for nucleotide binding and orientation. They suggest that spontaneously synthesised short peptides in the primordial environment may have served as ordered templates for the assembly of nucleotides. There would be selective advantage for a chemoton to enhance such an assembly process if the peptide conferred an advantageous feature.

It seems, therefore, that logical scenarios can be suggested for the very early evolution of coordinated interaction between nucleic acids and peptides that might lead to something like a primitive protein synthesis machinery. It is then a relatively easier step to postulate selective advantage in creating and then maintaining a copy of the 'messenger-RNA' sequence in the form of more chemically stable DNA. Synergistic ('symbiotic') associations of this sort must have been important in primeval biology; certainly, much current biology is dependent on synergistic relationships between molecules and symbiotic relationships between organisms. Primitive Earth (including the pre-Earth before the loss of our reducing atmosphere) must have been, literally, a hotbed of chemical reactions with all sorts of chemical experiments taking place in all sorts of places, with opportunities for an enormous range of mixed reactions.

This reflects my view that on the primeval Earth all chemical pathways for which a consistent argument can be constructed now, were effective then. If they could happen, they did happen. Where different pathways coexisted and were coextensive they could work together. Where they were isolated they would proceed in isolation to create chemotons appropriate to their environment until the next volcanic explosion and/or hurricane lifted those chemotons or other pre-alive systems (see below) into atmospheric aerosols and the global turbulence of the Earth brought them together with different chemotons/pre-alive systems from different places.

It is not unusual to suggest that different primitive evolutionary streams proceed independently and then come together to produce a more advanced combination, but the scope is usually limited to two – and most often specifically the 'metabolic' and 'genetic' streams. Two examples will suffice:

> It is logically possible to postulate organisms that are composed of pure hardware and capable of metabolism but incapable of replication. It is also possible to postulate organisms that are composed of pure software and capable of replication but incapable of metabolism. And if the functions of life are separated in this fashion, it is to be expected that the latter type of organism will become an obligatory parasite upon the former ... The first protein creatures might have existed independently for a long time, eating and growing and gradually evolving a more and more efficient metabolic apparatus. The nucleic acid creatures must have been ... obligatory parasites from the start, preying upon the protein creatures and using the products of protein metabolism to achieve their own replication. (Dyson, 1999, chapter 1, pp. 9/10)

And:

> In fact, a chemoton is simply [a] proliferating microsphere into which a template polymerization subsystem (which also has self-reproducing properties) is incorporated. (Gánti, 2003, chapter 2, p. 44)

In contrast to these quotations, what I am suggesting here is that there were many parallel schemes evolving towards relatively limited objectives at the beginning and consequently many contributions to the origin of our last universal common ancestor. On the basis that different types of pre-alive systems (protocells), each good at one particular process, must have coexisted, I believe that some might have found selective value in collaborating. Selective advantage would also be found in symbiotically absorbing and using other protocells, and in exuding factors (primitive enzymes and/or toxins) that targeted other protocells to lyse them and make use of their constituents (the protocellular origin of the fungal 'recycling' lifestyle).

I have already mentioned several of the pre-alive systems that have been proposed, under various names, over the years (Chapter 9) and that I believe were swept up into the atmospheric aerosol clouds to become candidates to make contributions to LUCA:

(a) **Coacervates** were proposed by Oparin (1957a, 1957b) to be the starting point for evolution leading to the origin of life. Oparin suggested that different types of coacervates containing different combinations of organic molecules were formed and were then subjected to a selection process from which eventually a living system arose. Several organic molecules were thought of in the 1950s as candidates to form the membrane that surrounded coacervates (see Chapter 9).
(b) **Lipid vesicles** or **liposomes**; these form spontaneously in solutions of fatty acids, lipids and phospholipids. They have come to epitomise coacervates because of the close similarity of the liposome membrane with the present-day cell membrane, and because of the ease with which they arise from chemicals that are reasonably expected to be present on

the primeval Earth (see Chapter 9; Deamer, 1985; Deamer & Weber, 2010; Deamer *et al.*, 2002).

(c) **Jeewanus** (a Sanskrit word meaning 'particle of life') are microscopic spherules that form membranes and also have some metabolic activity, including a primitive photosynthetic ability and ability to fix nitrogen. They spontaneous form in chemical mixtures made up of paraformaldehyde with colloidal molybdenum oxide as catalyst, or paraformaldehyde and potassium nitrate, in the presence of ferric chloride as a catalyst. When exposed to sunlight, about a dozen amino acids are synthesised, apparently by fixing atmospheric nitrogen (Bahadur, Ranganayaki & Santamaria, 1958).

(d) **Organic microstructures** are observed in the yellow polymeric accumulations in Miller–Urey spark discharge flasks. There are three types: 20×40 μm spheroids with textured surface and a multilayered membranous interior; 10×15 μm spheroids with a few surface folds and a complex interior composed of 0.03 μm granules embedded in a membranous matrix; and 1×2 μm rods or balls composed of 0.03 μm granules dispersed on a membranous matrix. The microstructures demonstrate autocatalytic, energy-dependent assembly (Fraser & Folsome, 1975).

(e) **Proteinoid microspheres** emerge from theoretical considerations when the ease with which amino acids form spontaneously is compared to the ease with which they polymerise, combined with analysis of extraterrestrial samples (Moon rocks, meteorites, sterile volcanic surfaces, etc.) and the results of Miller–Urey and similar experiments. The implications of similarities in the range and concentration patterns of amino acids, and the types of and structures of peptides found in all these sources are the inferences that: 'The fact that amino acids are self-instructing during polymerization explains the absence of a need for nucleic acids for the first proteins. Enzymes were thus available for a multiplicity of functions including the production of the first nucleic acids in their coding function ... The stages of prebiotic precursors to life and the following ones of protolife and life itself would have conformed both to the second law [of thermodynamics] and evolutionary theory' (quotation from Fox, 1980). The author concludes that 'The various stages in inanimate matter, protocells, and evolved cells and the degree of order that they represent comport with the second law of thermodynamics on a cosmic scale' (Fox, 1980).

(f) **Chemotons** are abstract minimal models for living organisms introduced by Tibor Gánti in 1971 (Gánti, 2003); they are fluid machines created from autocatalytic chemical cycles; see Fig. 9.3.

(g) **Surface metabolists** arising on the mineral surfaces of coprecipitates of FeS/NiS with other metal salts in and around deep, hot and acidic hydrothermal vents (black smokers). See Chapter 6 and Fig. 9.1 (Wächtershäuser, 2000, 2006).

(h) **Chemiosmotic bubbles** of iron sulphide with precipitated mineral membranes in cool alkaline water around shallow water hydrothermal vents (so-called 'off-axis' vents because they are some way away from the narrow neovolcanic zone of the volcanic ridge [the axis]; also known as white smokers); see Fig. 9.2 (Martin & Russell, 2007; Russell & Hall, 2002).

(i) **Mineral** and **crystal surfaces** (such as clays) might have been able to organise early metabolism as patterns of ionic charges generated by metal ions embedded in the regular silica crystal lattice specifically adsorb organic ('nutrient') molecules and catalyse specific chemical reactions. Further rounds of crystallisation would replicate the information contained in its ionic patterns, so the crystal serves as both catalyst of metabolism and store of genetic information (Cairns-Smith, 1982).

I am suggesting that rather than any one of these sources dominating the process, all of these options made a contribution to the origin of life (on the principle that if it could happen then it did happen). I would add the rider that the pre-alive systems that emerged from these different mechanisms were swept up into the aerosol clouds generated in the primordial atmosphere by storms and explosive volcanic activity so that all of these different contributions were brought together by atmospheric turbulence on a global scale. These interacted, and chemically reacted, in a new range of ways with chemical opportunities different from those at their point of origin and now under the additional influences of evaporative concentration, solar radiation, lightning discharges, a new range of gases in the atmosphere and volcanic plumes, and freezing and thawing cycles as they journeyed through the full height of the atmosphere.

Eventually the results of all these reactions were brought together in the rainwater and seawater trickling through the roofs of volcanic caves on the spindrift-washed shore of a volcanic island in an endless shallow ocean – where they mixed and interacted in yet another environment: namely the fluid films on the rocks within the caves, protected from the worst effects of the solar radiation and the dispersive effects of general wind and weather. So here is the field of play of the life game we sought at the end of Chapter 8: the trickling waters inside a cave, the dripping stalactites and the tiny gentle pools beneath.

Unlikely? Well of course it's unlikely; but even the most unlikely event will happen if there are enough opportunities for the event to occur. The probability of a UK National Lottery Lotto jackpot win has odds of 1 in 13 983 816; which is rather daunting for the individual lottery punter. But the fact is that in most weeks *somebody* wins. Next week it could be you! The problem with most origin of life stories is that they are based on the assumption that there is only one opportunity: on Earth, and in the ocean. Not in my story. Remember the estimate (Chapter 7) that the number of life-generating opportunities in aerosol reaction-vessels was in the region of 10^{31} (Donaldson *et al.*, 2002)? Well, we know it happened (otherwise we wouldn't be here writing/reading this), so that puts the odds at 1 in 10^{31}; which boils down to 'unlikely, but do-able'.

Once it emerged from its volcanic cave, LUCA could have become widely distributed. Fenchel (2002) shows this calculation:

> If the first cell (with a volume of 1 μm^3) was capable of a division every 24 hours, and there were large amounts of unused resources, then it would take only a little more than 100 days for the descendants to cover the Earth's surface with a 1 cm thick layer. (Fenchel, 2002, chapter 2, p. 8)

LUCA must have diversified rapidly, invading all the habitats on offer on a previously sterile world. Remember that these were anaerobic habitats and as prebiotically synthesised supplies of reduced carbon diminished, selection pressure would have favoured anaerobic autotrophic metabolisms. Canfield, Rosing & Bjerrum (2006) review the most probable electron donors and electron acceptors available to early-Earth (marine) ecosystems and conclude that the most active were probably driven by hydrogen + ferrous ion cycling, with microbial cycling of sulphur and nitrogen also feasible. The marine (and in some models deep oceanic) environment is given prominence because of the widely held view that the UV-radiation environment of the early Earth was too hostile for unprotected life forms at the surface. However, Westall *et al.* (2006) identify a 3.5–3.3 billion year old shallow sediment formation from South Africa as a 'well-developed microbial mat that formed on the surface of volcanic littoral sediments in an evaporitic environment'. They further suggest that the mat was constructed under flowing water, and included abundant extracellular polymers, filamentous anoxygenic photosynthesisers and rod-like sulphur-reducing bacteria. Cracks in the surface of the mat and the presence of minerals known to form by evaporation from aqueous solution indicate that the mat was periodically exposed to the air. These observations take us away from planktonic microbes in the open ocean and back to that volcanic cave on the spindrift-washed shore of a volcanic island; though now we see it some time later than before and our attention is drawn to the banks of its drainage channels and outflowing streams. The observations also imply that early organisms had efficient mutation repair and survival strategies, and/or:

> ... that DNA-damaging UV radiation fluxes at the surface of the Earth at this period must either have been low (absorbed by CO_2, H_2O, a thin organic haze from photo-dissociated CH_4, or SO_2 from volcanic

> outgassing; scattered by volcanic, and periodically, meteorite dust, as well as by the upper layers of the microbial mat ... (Westall *et al.*, 2006)

The observation of such an ancient microbial mat is extremely significant in view of the importance of biofilms in present-day microbiology. The morphological, chemical and other characteristics of the 3.4 billion year old mat components are the same as those of modern microbial biofilms. Present-day biofilms are microbial communities that live attached to surfaces in a matrix composed of mixed microbial cells, polysaccharides and excreted cellular products, including enzyme activities; they have been the subject of an enormous amount of research over the last decade or so (Sutherland, 2001). Practically all types of bacteria can form biofilms and this may be the preferred form of growth of bacteria in their natural environments (Karatan & Watnick, 2009). Indeed, living in biofilms has been described as 'the oldest, most successful and ubiquitous form of life' (Flemming, Wingender & Szewzyk, 2011) and 'the most successful forms of life on earth' (Flemming & Wingender, 2010). Given that present-day biofilms are both ubiquitous and successful it is reasonable to assume that the lifestyle is an ancient one; perhaps even the original one.

> Although biofilms are commonly referred to as 'slime', which implies that they are not rigid structures, their mechanical stability is important. Interestingly, it seems to be mainly the exopolysaccharides in the matrix that provide this feature. (Flemming & Wingender, 2010, p. 631)

If LUCA emerged in a biofilm we have to understand the advantages of the ancient biofilm for the origin of life. The only way we can do that is to look at present-day biofilms and try to assess what aspects of these biofilms could have provided opportunities for evolutionary advancement in the primeval environment.

The present-day world is dominated by microorganisms but they are microscopic and therefore essentially invisible to us in our everyday life, so our image of them is heavily influenced by what we know of their behaviour in the laboratory, either by direct experience or through the medium of TV films or educational videos. If I say 'bacterial population' you probably form a mental image of a volume of fluid with tens of millions of tiny cells floating around in it. That's certainly how many of them are cultured in the laboratory, but few of them live like that outside the laboratory in the natural world. Instead of living as pure cultures of dispersed single cells, microorganisms in nature accumulate as mixed populations at interfaces; that is, where at least two different substrata meet – such as the water/atmosphere interface on the surface of aqueous fluids (or in the foam that might be on the surface of agitated fluids), or the fluid/solid interface between a trickle of water and the rock beneath. At these interfaces whole communities of microbes gather as films, mats, flocs, or sludge and these are the different kinds of 'biofilm', which is the general name given to a thin coating composed of living material.

Biofilms form on all sorts of surfaces in the natural environment; for example, rocks and minerals, and man-made structures like stonework, brickwork, steelwork, concrete structures, and even the horn, bones and exoskeletons of dead organisms. They also form on inside and outside surfaces of plants, other microbes, and animals. In fact, even suspended aggregates of cells display many of the characteristics that are associated with biofilms. Essentially, wherever bacteria occur in our present-day world, they occur in biofilms; there are as many different types of biofilm as there are bacteria, and one species may make several different types of biofilms under different environmental conditions. Morphologically, biofilms can be smooth and flat, rough, fluffy or filamentous, and they may or may not contain water-filled voids. The biofilm matrix is composed of a range of extracellular polymeric substances (EPS), but the principal component of the matrix is water. Many EPS are hygroscopic and actively retain water, so the biofilm matrix provides a highly hydrated environment that dries more slowly than its surroundings and consequently protects cells in the biofilm against fluctuations in water availability.

> Desiccation seems to be one of the environmental conditions under which EPS provides global benefits to both EPS producers and other members of the biofilm community ... (Flemming & Wingender, 2010, p. 632)

Biofilm research is important in medicine because formation of a surface biofilm (often internally in the body) is a frequent first step in the attack of a pathogen; but biofilms are often involved in nosocomial infections (originating in a hospital) because they can form on hospital surfaces and medical equipment, even prosthetic devices and catheters. It is also important in biotechnology; on the one hand biofilms may be essential to a particular process (biofilms recycle your sewage), and on the other hand chemical engineers can spend sleepless nights trying to prevent biofilm formation in processes where uniform growth in suspension is most important for efficient formation of the desired product.

Bacteria may attach to surfaces as single cells or as clusters of cells; single cells form monolayer biofilms, where individual bacterial cells are attached only to the surface and cell-to-surface interactions rather than cell-to-cell interactions predominate. Bacteria attaching as clusters of cells form a multilayer biofilm; in this case each bacterial cell is attached both to the surface and to neighbouring bacteria and cell-to-cell interactions predominate. Monolayer biofilms may aggregate into multilayer biofilms, and may be the more pervasive surface-attached state in both the natural environment and the interaction of bacterial pathogens with their hosts. In many environments, the surface characteristics (meaning the pattern of electrostatic charges on extracellular proteins and/or polysaccharides) of bacteria lead to repulsion from common surfaces, so they need a mechanism to overcome this initial repulsion. Certainly, in motile bacteria flagellum motion accelerates surface adhesion for many bacteria by enhancing initial collisions with the surface by enabling the bacterium to overcome long-range repulsive forces, and so increasing the likelihood of close approach. The flagellar arrest that follows initial contact may then prepare the bacterium for later progression to multilayer biofilm development.

Pili are another type of long appendage that might occur on the surface of bacterial cells. Some pili are retractable, and retractable pili are a common requirement for attachment of gram-negative bacteria to surfaces in the present day. This also seems to be a matter of overcoming initial repulsion between surface and cell. The extended pilus tip can attach to the surface and then retract, pulling the bacterial cell either onto or along surfaces. In many bacteria, a retractable pilus and/or flagellum is only able to achieve a transient attachment that may be disrupted; the attachment needs to be stabilised by the cell producing specific 'adhesins', which are usually large secreted proteins or polysaccharides that make the attachment permanent. Multilayer biofilms are usually formed with an extensive extracellular matrix that includes polysaccharides, proteins and DNA. Indeed in most biofilms, the microbial cells account for less than 10% of the dry mass, whereas the matrix accounts for over 90%. This extracellular matrix is made up of different types of extracellular polymeric substances (EPS) produced by the organisms themselves; such large quantities are produced that the EPS make up a major fraction of available carbon in soils, sediments, and suspended aggregates of oceanic and fresh water. The matrix forms the scaffold for the three-dimensional architecture of the biofilm within which the cells are embedded. The matrix is responsible for adhesion to surfaces and for cohesion of the biofilm, and it is the matrix that allows the embedded cells to enjoy a lifestyle that is entirely different from the planktonic state (Flemming & Wingender, 2010; Karatan & Watnick, 2009). In particular:

> ... keeping them in long-term close proximity and, thus, allowing intense interactions to occur, including cell–cell communication, horizontal gene transfer and the formation of synergistic microconsortia. (Flemming & Wingender, 2010, p. 623)

Several functions of extracellular polymeric substances have been established (Table 10.1) and demonstrate the wide range of advantages for the biofilm mode of life in the present day, many of which probably applied to ancient biofilms (Fig. 10.1). The major components of the matrix – proteins, polysaccharides, lipids and perhaps most interesting of all, DNA – are not

Table 10.1. Functions of extracellular polymeric substances in bacterial biofilms of the present day

FUNCTION	RELEVANCE FOR BIOFILMS	EPS COMPONENTS INVOLVED
Adhesion	Allows the initial steps in the colonisation of abiotic and biotic surfaces by planktonic cells, and the long-term attachment of whole biofilms to surfaces	Polysaccharides, proteins, DNA and amphiphilic molecules
Aggregation of bacterial cells	Enables bridging between cells, the temporary immobilisation of bacterial populations, the development of high cell densities and cell-to-cell recognition	Polysaccharides, proteins and DNA
Organisation of biofilms	Forms a hydrated polymer network (the biofilm matrix), mediating the mechanical stability of biofilms (often in conjunction with multivalent cations) and, through the EPS structure (capsule, slime or sheath), determining biofilm architecture, as well as allowing cell–cell communication	Neutral and charged polysaccharides, proteins (such as amyloids and lectins) and DNA
Binding of enzymes	Allows accumulation, retention and stabilisation of enzymes through their interaction with polysaccharides in the matrix	Polysaccharides and enzymes
Enzymatic activity	Provides a cell-controlled local environment for enzyme activity. Enables the digestion of exogenous nutrients the cells can then acquire; also allows for the degradation of structural EPS, allowing the release of cells from biofilms	Proteins
Protective barrier	Confers resistance to non-specific and specific host defences during infection, and confers tolerance to various antimicrobial agents (for example, disinfectants and antibiotics), as well as protecting cyanobacterial nitrogenase from the harmful effects of oxygen and protecting against some grazing protozoa	Polysaccharides and proteins
Retention of water	Maintains a highly hydrated microenvironment around biofilm organisms, leading to their tolerance of desiccation if water supply unreliable	Hydrophilic polysaccharides and, possibly, proteins

Take up (sorption) of organic compounds	Allows the accumulation of nutrients from the environment and the sorption of drugs, pesticides or carcinogens (xenobiotics), thus contributing to environmental detoxification [sorption includes the processes of absorption and adsorption]	Charged or hydrophobic polysaccharides and proteins
Sorption of inorganic ions	Promotes polysaccharide gel formation, ion exchange, mineral formation and the accumulation of toxic metal ions (thus contributing to environmental detoxification)	Charged polysaccharides and proteins, including inorganic substituents such as phosphate and sulphate
Export of cell components	Releases cellular material as a result of metabolic turnover	Membrane vesicles containing nucleic acids, enzymes, lipopolysaccharides and phospholipids
Nutrient source	Provides a source of (reduced) carbon-, (fixed) nitrogen-, phosphorus- and sulphur-containing compounds for utilisation by the biofilm community	Potentially all EPS components
Exchange of genetic information	Facilitates horizontal gene transfer between biofilm cells (by keeping cells in close proximity for long periods)	DNA
Electron donor or acceptor	Permits redox activity in the biofilm matrix (transfer of electrons such that one reactant is oxidised by loss of electrons, while another is reduced by gain of electrons)	Proteins (for example, those forming pili) and, possibly, humic substances (insoluble organic constituents of soil)
Sink for excess energy	Stores excess carbon if the ratio of available carbon to nitrogen becomes unbalanced	Polysaccharides

EPS, extracellular polymeric substances.
Table modified from Flemming & Wingender (2010).

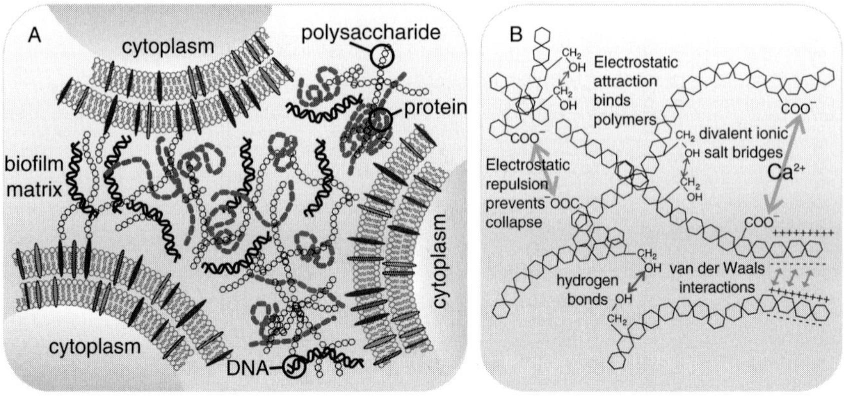

Fig. 10.1 The biofilm matrix. **A**, a cartoon drawing of the major polymeric components of the present-day biofilm matrix, polysaccharides, proteins and DNA, illustrated in a segment of matrix surrounded by the boundaries of three negibacteria, with cells bounded by two acyl ester phospholipid membranes. **B**, indicates the principal weak physicochemical interactions and entanglements of biopolymers that dominate the stability of the present-day biofilm matrix and which, as fundamental chemical interactions, would have existed in the primeval biofilm of the last universal common ancestor (LUCA) and the various pre-alive systems (coacervates, liposomes, jeewanus, organic microstructures, proteinoid microspheres, chemotons, surface metabolists on FeS/NiS coprecipitates, chemiosmotic bubbles of iron sulphide, clay and other crystal surfaces) from which LUCA evolved. (Redrawn and adapted from Flemming & Wingender, 2010.)

distributed homogeneously between the cells, so they set up differences between regions of the matrix. Microenvironments are essentially constructed by the cells they contain and within which the cells they contain can get on with living in their local environment the way they want to live. These components:

> ... provide the mechanical stability of biofilms, mediate their adhesion to surfaces and form a cohesive, three-dimensional polymer network that interconnects and transiently immobilizes biofilm cells. In addition, the biofilm matrix acts as an external digestive system by keeping extracellular enzymes close to the cells, enabling them to metabolize dissolved, colloidal and solid biopolymers. (Flemming & Wingender, 2010, abstract)

The protein content of the biofilm matrix can greatly exceed the polysaccharide content. A variety of extracellular enzymes can be efficiently retained in the biofilm matrix by their interaction with polysaccharides. Several activities have been detected, many of them being involved in degradation of biopolymers, both water-soluble polymers (many polysaccharides, proteins and nucleic acids) and water-insoluble polymers (cellulose, chitin and lipids). These enzymes turn the matrix into an extracellular digestive system able to break down biopolymers to their monomers that can then be taken up and utilised as nutrients and energy sources. Breakdown of structural EPS will also assist in detaching bacteria from biofilms. Non-enzymatic proteins in the matrix include carbohydrate-binding proteins (called lectins) that create and stabilise the matrix network of polysaccharides and make bonds between the bacterial surface (wall and membrane) and the extracellular EPS.

Many different polysaccharides can be found in biofilms. Several are homopolysaccharides (made up of a single type of sugar), including cellulose and other glucans, fructans (fructose polymers) and mannans (mannose polymers). Most biofilm polysaccharides, though, contain a mixture of neutral and charged sugar residues (i.e. heteropolysaccharides) and can contain chemical groups that greatly affect their physical and biological properties.

Uronic acids – sugars in which the terminal carbon–hydroxyl function (–CHOH) has been oxidised to a carboxylic acid (= COOH) – form polyanionic polysaccharides (having many negative charges), including alginate (which contains mannuronate and guluronate) and xanthan (contains glucuronate). Polycationic polysaccharides (having many positive charges) also exist in the biofilm matrix and feature *N*-acetylglucosamine residues. Acetyl groups commonly replace hydrogen atoms in biofilm polysaccharides as they increase the adhesive properties and interconnections within the EPS and alter biofilm architecture. Biofilm architecture is also strongly influenced by interaction of anionic EPS (containing carboxylic acid groups) with multivalent cations, especially Ca^{2+} and other divalent metal ions.

> The EPS matrix can act as a molecular sieve, sequestering cations, anions, apolar compounds and particles from the water phase. EPS contain apolar regions, groups with hydrogen-bonding potential, anionic groups (in uronic acids and proteins) and cationic groups (for example, in amino sugars). (Flemming & Wingender, 2010, p. 631)

Many present-day biofilms of various origins have been found to contain extracellular DNA (eDNA). Some of this is residual genomic material from dead and burst cells, but not all. At least some of the eDNA is non-genomic and specially produced, and excreted from live cells, as an essential structural part of the biofilm matrix where it contributes to the mode of life by enabling the aggregation of bacteria (the process is technically called flocculation). In other cases eDNA functions as a cell-to-cell connector or as an adhesin, fixing the biofilm to surfaces. Some eDNA also has antimicrobial activity, attacking the outer membrane of sensitive cells by selectively removing the positively charged ions (cations) that stabilise the bacterial membrane (this process is called chelation; the eDNA has a higher affinity for the cations than the membrane components). Biofilms in waste-water treatment plants have particularly large amounts of eDNA but the use of eDNA by different species of bacteria is highly variable; for example, eDNA is a major structural component in the biofilm matrix of *Staphylococcus aureus*, but only a minor component of biofilms formed by *Staphylococcus epidermidis* (where biofilm structural integrity depended on the polysaccharide poly-*N*-acetylglucosamine (PNAG), which has no such function in *S. aureus*) (Flemming & Wingender, 2010).

The polysaccharides, proteins and DNA of the biofilm matrix are hydrophilic and highly hydrated molecules; but there are interactions that would be better served by molecules with hydrophobic properties. This is where the surfactants and lipid components of the EPS come into prominence. Polysaccharide-linked methyl and acetyl groups and lipopolysaccharides are crucial for the surface-active properties of EPS. EPS with hydrophobic properties enable the biofilm to adhere to waxy, oily, and in the present day, to plastic (including Teflon) surfaces. This last is particularly important in the clinical environment because it is the hydrophobic EPS that enable biofilms to build up on medical devices and equipment. In the natural environment of today the surfaces of plants and of insect exoskeletons are waxy, so biofilms of plant and insect pathogens and epiphytes need hydrophobic adhesins.

What sort of picture of the first (primeval) biofilm can we extract from these observations of present-day biofilms? The nature of the EPS making up the matrix of the first biofilms is likely to have reflected the various polymers in the local environment resulting from non-biological synthesis, including PAHs, peptides, lipids, etc. (see Chapters 6 and 7). There were other sources, in addition:

(a) aerosol droplets acting as chemical reaction vessels but failing to reach the successful conclusion of a pre-alive system will be rained out of the atmosphere to contribute whatever organic molecules they had made to the biofilm;
(b) partially successful pre-alive systems of various sorts (coacervates, liposomes, jeewanus, organic microstructures, proteinoid microspheres, chemotons, surface metabolists on FeS/NiS coprecipitates, chemiosmotic bubbles of iron sulphide, clay and other crystals) from which LUCA evolved would be supported within the biofilm until they reached the end of their life, at which time their component organic molecules would be contributed to the biofilm matrix;
(c) as more successful pre-alive systems 'learned' to make polymers (polysaccharides, peptides, nucleic acids, lipids), it is likely that their primitive boundary membranes would be relatively unspecifically permeable,

leading to leakage of polymers into the biofilm matrix. Simulations of competition in present-day biofilms show that polymer producers have adaptive value over non-producers, and this probably also applied in primeval biofilms.

Biofilm existence would have offered several selective advantages for evolution of the first cells from pre-alive systems. Clearly, physical containment within a biofilm would have been of benefit at the earliest stages by helping the first cells to maintain their structural integrity and providing protection from desiccation and solar radiation. Reduced mixing and dispersion (the equivalent of the present-day segregation of microdomains within biofilms) would also have been important as a means of bringing together and keeping together the rare successful pre-alive systems so they could interact. It is likely that competition for nutrients and co-operation within the EPS matrix would lead to constant adaptation towards greater fitness. As is the case today, the primeval biofilm matrix would:

(a) accumulate early enzymes to limit their loss by dispersion;
(b) sequester dissolved and particulate nutrients from the general environment;
(c) act as a location for recycling the components of dead pre-alive systems;
(d) eventually, serve as a nutrient source.

In addition to all these considerations there are a number of intriguing possibilities for the ancient biofilm matrix offering initial positive selection pressure for features that later achieved adaptive value in other circumstances. For example, the biofilm matrix could be the place where DNA comes into the origin of life story, but simply as a structural component of the matrix. It is easy to imagine that after the ability evolved to synthesise structural DNA (which has adaptive value because it is 'almost a polysaccharide') for the biofilm matrix, other aspects of DNA function could show their adaptive value. For example, the greater stability of the poly-*deoxy*-ribonucleotide in comparison with the poly-ribonucleotide would clearly have sufficiently decisive adaptive value to use DNA as a matrix structural component, for DNA to be selected for that role irrespective of any value it may later express as a genetic molecule. Later, though, as a genetic molecule, that same chemical stability is also important to its potential function as a store of genetic information. Similarly, the metabolite binding and chemical catalytic abilities of RNA, coupled with its relative instability (which makes it controllable), would make RNA suitable for metabolic functions as a ribozyme irrespective of any value it may later express as a genetic adapter molecule enabling one language (DNA genetic code) to be translated into another (protein amino acid sequence).

The ability to synthesise and externalise polysaccharides for the biofilm matrix is another feature that will have selective advantage for the biofilm matrix irrespective of adaptive advantages that may be expressed later in evolution by specific polymer molecules. The polysaccharide functions required for the matrix can be served by a wide range of polymers; consequently, there is greater selective value in making *any* sugar polymer than in making any *one particular* polymer; even though today's polymers like cellulose and chitin have very particular functions far removed from those needed in the biofilm matrix. Controlled synthesis is not even required; any chemoton that can stitch together sugars into a polymer and export the product to the environment will be a welcome member of the biofilm community. Regulation, control and precision will follow in time.

Yet once the reliable synthesis and export of polysaccharides has been established to serve the purposes of the biofilm matrix the molecules will be available for further rounds of selection of other functions. Similarly, being able to synthesise polysaccharides for the biofilm matrix makes those molecules available for chemical modification and, remember, there is an ancient ability to acetylate in the primeval chemistry of the iron-sulphur world (Chapter 9). Initially, just because the substrates are confined together, acetylation will occur; randomly at first, but in an increasingly directed manner as the products of acetylation reveal new functions of value to the matrix and the pre-alive systems it contains.

Here, then, is the field of play for the life game: a film of water, perhaps no more than a few tens of micrometres deep. It is covering the roof and walls of some of those bubbles of magma that were burst by the explosion of steam when the lava reached the sea just a few years before. Now warm, rather than scaldingly hot; the film of water in the cave-like magma bubble is protected from solar radiation, but is open to the atmosphere. It dries, and its solutes crystallise out; it is replenished by the rain from the volcanic plume above, then dries again, only to be once more replenished as storm-driven spindrift soaks the shore. And all the time this film of water is gathering together the harvest of aerosol-contained chemical reactions from the global atmosphere and those of interplanetary and interstellar origins in the chondrites and extraterrestrial dust showering down from space. Gradually, the highly polymerised molecules accumulate (PAHs from space, overcooked Miller–Urey and iron-sulphur reactions from the atmosphere and volcanic plumes) and the water film becomes a film of slime – to which a constant aerosol rain adds organic compounds of every type, as well as coacervates, liposomes, jeewanus, organic microstructures, proteinoid microspheres, chemotons, surface metabolists on FeS/NiS coprecipitates, chemiosmotic bubbles of iron sulphide, clay and other crystals and microscopic aggregates. Thus, the slime film becomes a primeval biofilm in which:

(a) chemotons compete for substrates, and merge and adapt as new substrates become available;

(b) liposomes absorb chemiosmotic bubbles to make themselves into ion-pumping vesicles, and jeewanus components to experiment with photochemistry;
(c) proteinoids wrap themselves around FeS/NiS coprecipitates and microscopic crystals to discover organometallic chemistry;
(d) coacervates and liposomes enlarge by adding components to their membranes and absorb a more diverse range of the organic microstructures that surround them, and by so doing create a protected microenvironment in which the community of microstructures can flourish, interact, integrate, synergise;
(e) coacervates and liposomes containing different communities of microstructures compete, flourish, interact, merge, integrate, synergise.

And LUCA emerges from the long night that started when the Sun first set alight.

You can forget all those theories in which the origin of life occurs in an oceanic primeval soup or in a deep, hot place somewhere, or even a warm little pond. Life originated as a biofilm, and the precursors and components of the first fully working biofilm were brought together by octillions of drifting aerosol droplets from around the globe that acted as dynamic reaction vessels. You don't need to stare dreamily into the distant sky, or to the far horizon of the boundless ocean to find the field of play where life's game on Earth was first played out. It's in the rainwater and seawater trickling through the roofs of volcanic caves on the spindrift-washed shore of a volcanic island in an endless shallow sea. Step carefully; it's in the slime on the volcanic sand at your feet.

ELEVEN

TOWARDS EUKARYOTES

Prokaryotes have dominated the Earth for the bulk of its history (I have put that statement in a tense that suggests they do not dominate the Earth now, but the truth might be other than this). LUCA must have emerged close to the start of the Archaean Eon, about 3.8 billion years ago, because, as noted above, some of the oldest microbial fossils are fully differentiated, photosynthetic bacteria (cyanobacteria) found in Western Australian sediments that are 3.5×10^9 years old (Boal & Ng, 2010; Derenne *et al.*, 2008; Schopf, 1993). By contrast, eukaryotes are generally thought to have appeared no earlier than about 1.5 billion years ago (and some people put their emergence somewhat later than that). So, for at least 2 billion years the only living organisms on the planet were prokaryotes together, presumably, with their associated viruses.

On the basis of protein sequence comparisons, LUCA probably had a complexity comparable to that of a simple modern bacterium and lived 3.2–3.8 billion years ago (Orgel, 1998). The abundant biological activity in the deep ocean volcanic hydrothermal systems of the present day, most of it being dependent on chemosynthesis rather than photosynthesis, has stimulated the widespread appeal of the theory promoting a 'deep hot' origin of life, and particularly Günter Wächtershäuser's argument linking the chemistry of submarine deep ocean vents with the origin of life (Alpermann *et al.*, 2010; Wächtershäuser, 2006):

> Wächtershäuser asserts that life originated on the surface of iron sulfides as a result of such chemistry. The assumptions that complex metabolic

> cycles self-organize on the surface and that the significant products never escape from the surface are essential parts of this theory; in Wächtershäuser's opinion, there never was a prebiotic soup! (Orgel, 1998)

The chemistry is plausible and the notion suggests a prebiotic biofilm. However, a chemoautotrophic origin of life in a deep volcanic vent iron-sulphur world implies that the pioneer organisms were hyperthermophiles. In the present day:

> ... hyperthermophilic ('superheat-loving') bacteria and archaea are found within high-temperature environments, representing the upper temperature border of life. They grow optimally above 80°C and exhibit an upper temperature border of growth up to 113°C ... (Stetter, 2006, abstract)

Hyperthermophiles currently inhabit anaerobic environments where they exhibit chemolithoautotrophic nutrition, gaining energy by inorganic redox reactions involving nutrients like molecular hydrogen, carbon dioxide, sulphur, and ferric and ferrous iron. Karl Stetter discusses the phylogenetically earliest lineage of archaea, which is represented by the extremely tiny members of the kingdom Nanoarchaeota that thrive in submarine hot vents (Stetter, 2006). This group occupies the short, deep branches of the phylogenetic tree of life that, on the basis of gene-sequence comparisons, the computer programs place closest to the root, suggesting they might have existed on the early Earth; it is this that provides evidence that LUCA might have been a hyperthermophile.

Emerging from these arguments we have what might be called a conventional, or 'textbook', phylogenetic tree of life. This is not without criticism, though. I have already quoted Lazcano & Miller (1996) on the topic ('Hyperthermophiles may be cladistically ancient, but they are hardly primitive'); the reason for this comment is that hyperthermophiles have such an extensive catalogue of unique molecular characteristics that it is difficult to believe that they can all be ancestral rather than highly adapted. For example, archaean hyperthermophiles have several tungsten-containing enzymes, unusual heat shock proteins, a reverse gyrase that sets up positive supercoils in DNA to improve the temperature stability of genomic DNA, and key enzymes of glycolysis (phosphofructokinase and hexokinase) that require adenosine ***di***phosphate (ADP) rather than the more usual adenosine ***tri***phosphate (ATP), which is unstable at high temperature (Robb & Clark, 1999). Furthermore, intrinsically heat-stable proteins are typical of hyperthermophiles and the structural stability of many thermophile proteins in the 100–130 °C range is achieved by several chemical processing events that the proteins experience after their synthesis (known as post-translational modification). Proteins can be modified by permanent chemical attachment of several types of molecule. Examples of such modifications include (Eichler & Adams, 2005):

(a) The attachment of sugars or lipids, phosphorylation, and methylation.
(b) In most proteins disulphide bridges between closely positioned pairs of cysteine amino acids control the three-dimensional shape of a polypeptide chain, and can also be used to keep proteins in multisubunit complexes. Hyperthermophiles use a process of thiol-disulphide exchange reactions (known as disulphide shuffling) to change the location of disulphide bonds rapidly within a protein (or between proteins in a multiprotein complex) after the proteins have been synthesised.
(c) A hydrophobic core is a feature of stable proteins in less extreme hyperthermophiles than in those growing at or above 100 °C; the amino acids are more stable at elevated temperatures when located in such a hydrophobic core than in free solution.
(d) Networks of acidic and basic side chains on the surface of hyperthermophile proteins interact to form ionic networks that bind protein assemblies together and are more often observed in the proteins of hyperthermophiles than in homologous proteins in mesophiles.
(e) Many proteins that are monomeric in mesophiles are found to form oligomers in hyperthermophiles.

All of these modifications allow the cell to control the folding and function of a protein because loss in integrity of a temperature-sensitive protein molecule results from unfolding of the monomeric protein into the denatured form. Strengthened intramolecular and intermolecular associations like those provided by the modifications listed above help prevent unfolding.

> Indeed, archaeal proteins are able to remain properly folded and functional in the face of extremes of salinity, temperature, and other adverse physical conditions that would normally lead to protein denaturation, loss of solubility, and aggregation ... (Eichler & Adams, 2005, p. 394)

In mesophiles the cell membranes are made mostly of phospholipids arranged in a double layer with the hydrophobic tails from both layers facing toward each other and the hydrophilic heads facing outwards toward the aqueous environment at both surfaces. The general class of lipids, which includes fats and oils, are made from two kinds of molecule: glycerol, a three-carbon alcohol with hydroxyl groups on each carbon atom, and three fatty acids; each fatty acid is a long hydrocarbon chain, making it hydrophobic. These are known as triglycerides because three fatty acids are attached to the glycerol. Phospholipids are diglycerides made from glycerol, two fatty acids, and a phosphate group in place of the third fatty acid (the phosphate often has another molecule attached to it). The hydrocarbon tails of fatty acids in phospholipids are still hydrophobic, but the phosphate group end of the molecule is hydrophilic because of the unshared electrons of the oxygen atoms. This results in phospholipids being soluble in both water and oil (amphiphilic) and enables them to take up (spontaneously) the double layer structure of the typical cell membrane. Typical, that is, of all but archaea.

Archaea mainly use isopranyl ether lipids, which have branched hydrocarbon chains, instead of the normal phospholipids and fatty acids; and it's not just a minor addition. In *Pyrococcus woesei*, for example, the diether 2,3-di-O-phytanyl-sn-glycero-1-phosphoryl-19-myo-L-inositol comprises 90% of the lipids and there are no fatty acids. In ether lipids the carbon atoms of glycerol are bonded to the branched hydrocarbon chain through an ether linkage (≡C–O–C≡), as opposed to the usual ester linkage (–COOC≡) (Carballeira *et al.*, 1997). The branched ether lipids give hyperthermophiles improved membrane structure at temperatures that would disrupt normal phospholipid membranes.

All of these modifications involve quite sophisticated chemical alterations. They are indications of a highly adapted metabolism rather than an ancient and primitive metabolic chemistry. This is one reason why several people have sought an alternative origin of life under more temperate conditions than those offered by the deep, hot environments given credence by the conventional molecular phylogenetic trees. Another important reason (discussed above) is the discovery of fossils of very ancient microbial mats formed in shallow flowing water (Westall *et al.*, 2006); thus, in the last few years a completely different phylogeny has been suggested. By emphasising major innovations in cell structure, particularly bacterial cell envelope novelties and flagella, instead of relying on molecular cladistics, Cavalier-Smith (2006, 2010a) considers thermophiles to have evolved late; he has placed the Chlorobacteria as the most primitive prokaryote phylum, making Archaebacteria the youngest bacterial phylum and archaebacteria the sisters (rather than ancestors) of eukaryotes.

With the caveat that there is this alternative view (of Cavalier-Smith, 2006, 2010a; and to which I shall return), I will now outline the conventional story with a few quotations from a recently published textbook. What has become the standard method for drawing the tree of life is to build phylogenetic trees from genes that duplicated before the LUCA arose [rRNA is ribosomal-RNA]:

> The breakthrough was made by Carl Woese who concluded that, as all organisms possessed small subunit rRNA (SSU rRNA, so called because they form part of the small subunit of a ribosome), the small subunit rRNA gene would be a perfect candidate as the universal chronometer of all life. SSU rRNA genes (16S rRNA in prokaryotes and mitochondria, 18S rRNA in eukaryotes) display a mosaic of conservation patterns, with rapidly evolving regions interspersed among moderately or nearly invariant regions (it has been estimated that about 56% of the nucleotide positions in 18S rRNA data sets are not free to vary and have not

> undergone substitutions useful in phylogenetic reconstructions). This variation in conservation permits SSU rRNA gene sequences to be used as sophisticated chronometers of evolution with the slowly evolving regions recording events that occurred many millions of years ago, and the rapidly evolving regions chronicling more recent events.
>
> In Woese's procedure, pairs of SSU rRNA gene sequences from different organisms were aligned, and the differences counted and considered to be some measure of 'evolutionary distance' between the organisms. Pair-wise differences between many organisms were then used to infer phylogenetic trees, maps that represent the evolutionary paths leading to the SSU rRNA gene sequences of present-day organisms. Of course, such trees rely on many assumptions, among which are assumptions about the rate of mutational change (the 'evolutionary clock') and that rRNA genes are free from artefacts generated by convergent evolution or lateral gene transfer (Woese, 1987; Woese, Kandler & Wheels, 1990). (Moore *et al.*, 2011, p. 24)

Woese recognised three primary lines of evolutionary descent, first called kingdoms, but subsequently renamed domains as a new taxon above the level of kingdom; these three domains are now called Eubacteria, Archaea and Eukaryota. The first two domains contain prokaryotic microorganisms and the third domain contains all eukaryotic organisms. All were thought to have diverged from some universal ancestor, which is the ancestor of all organisms alive today; this is called 'rooting the tree of life' and the organism at the root of the tree of life is that which I have here called LUCA (Fig. 11.1).

The time of occurrence of evolutionary events in the tree of life cannot be extracted reliably from SSU ribosomal-RNA gene sequences alone, because the presumed evolutionary, or molecular, clock is not constant in different lineages (indeed, Cavalier-Smith, 2006, states categorically 'there are no real molecular clocks'). Other sequences have been brought into the analysis. Molecules chosen for phylogenetic studies must:

(a) be universally distributed across the group chosen for study;
(b) be functionally homologous;
(c) change in sequence at a rate proportionate with the evolutionary distance to be measured (the greater the phylogenetic distance being measured, the slower must be the rate at which the sequence changes).

The SSU ribosomal-RNA genes do satisfy these criteria, but single gene trees cannot be allowed to dominate the conclusions. ATPase genes (which specify the enzyme that hydrolyses phosphate from adenosine triphosphate) have been used in multigene phylogenies. ATPase enzymes are composed of several different kinds of subunits (called F-, V- and A-enzymes) all of which have catalytic and non-catalytic subunits. Using ATPase subunits, the root of

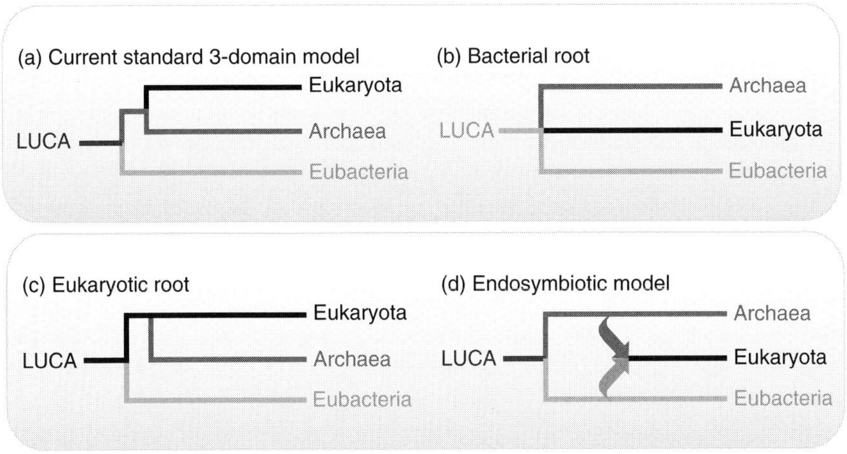

Fig. 11.1 Different ways of rooting the tree of life discussed in the text. What is here described as the 'standard' three-domain model based on Woese *et al.* (1990) is shown as diagram (**a**) and shows LUCA (the last universal common ancestor) diverging into bacterial and archaea + eukaryote stems; from the latter of which eukaryotes later emerge from their archaean ancestors. In (**b**) LUCA is a bacterial stem from which archaea and eukaryotes separately diverge; this makes Archaea a sister group to Eukaryota rather than its ancestor. The model which postulates that eukaryotes came first is shown in (**c**); here LUCA possessed basic eukaryotic cell biology and persisted as a stem eukaryote during the time that bacterial and archaean prokaryotes (which emerged by loss of features from the eukaryotic stem) dominated the biosphere. In (**d**) the endosymbiotic or syntrophic model proposes that the eukaryote stem arose when ancient archaea invaded eubacteria, the former evolving into the eukaryote nucleus, the latter providing the cell metabolism. All of these models also envisage that mitochondria and chloroplasts derived from endosymbiotic relationships between early eukaryotes and aerobic bacteria or cyanobacteria, respectively (not shown here). Also, all of these models are compatible with the ancestral stem eukaryote having a fungal body plan/lifestyle.

the tree of life was placed on the Eubacteria branch, making the Eukaryota and Archaea sister taxa. This conclusion was supported by studies using aminoacyl transfer-RNA synthetase gene sequences, which form a series of 20 enzyme families. Ribosomal genes are another regular contributor to construction of phylogenetic trees. The ribosomal genes are usually present in genomes as 100 to 200 tandem repeats, but they evolve as a single unit. Three regions of the ribosomal gene cluster in eukaryotes code for ribosomal-RNA genes, which are transcribed into 5.8S, 18S and 28S RNA molecules that form part of the ribosome structure. Of the approximately 9000 nucleotides in a ribosomal repeat, the 18S gene accounts for about 1800 base pairs, the 5.8S gene for about 120 base pairs and the 28S gene for about 3200 base pairs. Interspersed between the ribosomal-RNA genes are spacer regions. The areas that lie between the 18S and 5.8S and between the 5.8S and 28S genes are called internally transcribed spacers (ITS1 and ITS2). The region that separates one ribosomal gene cluster from the next is called the intergenic spacer region (IGS). The ribosomal-RNA genes and the different spacer regions all evolve at different rates and because of this their sequences have become widely used to discriminate between taxa at levels ranging from the kingdom to the intraspecific strains and races. Other sequences used in phylogenetic studies of eukaryotes include: ribosomal protein factors, α-tubulin, β-tubulin, actins and cytochromes. The consensus of these analyses was to suggest that LUCA was bacterial-like and that the root lay between eubacteria and (archaea plus eukaryota), with the archaea being ancestral to eukaryotes (Fig. 11.1a).

Thus, the molecular sequence data is strong, but is it strong enough? Inferring ancient relationships is full of difficulties, and analyses that share the same methods share the same difficulties. If the data contain too little information, random errors can swamp the truth, and random errors can be introduced by the mathematical model used to interpret the data. The method used to establish the phylogenetic trees can also introduce more systematic errors if it is too simplistic (explanations and further references in Keeling *et al.* 2005). Errors amplify as you attempt to reach back further in time, and we are interested in what is called deep time (hundreds of millions to billions of years ago) where we are looking for deep divergences between major groups of organisms. Evidence that genes could transfer laterally, which means between bacteria in the same generation, and that this occurred in ancient times, makes the distant origins of the major lineages of life uncertain.

The root of the universal tree of life remains controversial. Fossils can calibrate phylogenetic trees to a real timeline, but there is a very patchy fossil record in deep time and the older the fossil the greater the debate about its nature. Consequently, the timings inferred in different studies for major happenings can differ by several hundred million years. For example, one study claims the common ancestor of living eukaryotes existed 2.3 billion years ago; whilst another puts the time of eukaryote divergence at 0.95 to 1.26 billion years ago. The two studies used different methods and their different dates could mean that the common ancestor existed for a billion years before evolving into the fungal, plant, and animal lines, or it could mean that the best we can say is that the divergence event occurred somewhere between 0.95 and 2.3 billion years ago.

The associations inferred from molecular analyses can be satisfied by a number of root branching patterns (Fig. 11.1), and alternatives have been suggested:

> Penny & Poole (1999) pointed out that modern eukaryotes use RNAs to catalyse intron splicing and stable RNA processing and suggested that

> these processes could be 'molecular fossils' from the RNA world that may have been the first step in the origin of life, before the evolution of protein catalysis. So, the universal ancestor might have possessed some extremely primitive features that are now considered to be characteristic of present day eukaryotes. If these relics of the RNA world were present in the universal ancestor, it doesn't mean that the ancestor was eukaryotic. Rather, the ancestor contained a mix of features that were selected and combined in different ways during evolution of the present day archaea, eubacteria and eukaryotes. (Moore *et al.*, 2011, p. 24)

Indeed, Penny & Poole (1999) paint this picture of LUCA:

> It was a fully DNA and protein-based organism with extensive processing of RNA transcripts by RNPs [ribonucleoproteins]. It had an extensive set of proteins for DNA, RNA and protein synthesis, DNA repair, recombination, control systems for regulation of genes and cell division, chaperone proteins, and probably lacked operons. Biochemistry favours a mesophilic [preferring 30–40 °C] LUCA with eukaryote-like RNA processing, though it is still possible to fit the data to several different trees … (Penny & Poole, 1999, p. 676)

Unfortunately, gene trees are ambiguous and it is expecting a lot for the relationships between organisms that diverged 3–4 billion years ago to be determined unambiguously from small numbers of present-day genes. Phylogenetic analysis using the standard methods outlined above generally supports the 'eubacteria and (archaea plus eukaryota)' clustering. However, there are other methods that strongly suggest that this is an artefact because eubacteria evolve more rapidly than the other domains and ancient phylogenetic signals are lost in what is known as 'long- branch attraction'. These analyses provide support for a monophyletic grouping of prokaryotes (eubacteria plus archaea) and a eukaryotic root for the tree of life (Fig. 11.1); that is, LUCA was like the Penny & Poole (1999) description quoted above and eubacteria and archaea diverged from this as sister groups. There is still a high level of uncertainty, though (Brinkmann & Philippe, 1999; Lopez, Forterre & Philippe, 1999) and the root of the universal tree of life remains controversial.

A significant aspect of the controversy is the origin of the defining characteristic of the eukaryotic cell – its nucleus; eukaryotes have one, prokaryotes don't. Pennisi (2004) outlines the major theories that have been proposed to explain the origin of the nucleus:

(a) A symbiotic relationship between an archaean and a bacterium; this 'syntrophic or endosymbiotic model' proposes that ancient archaea, similar to modern methanogenic archaea, invaded and lived within bacteria similar to modern myxobacteria, metabolic synergism between the two brings them together and the archaean evolves into the eukaryote nucleus (Fig. 11.1d). This is analogous to the accepted theory for the origin of mitochondria and chloroplasts from a similar endosymbiotic relationship between early eukaryotes and aerobic bacteria or cyanobacteria, respectively (Margulis, 2004).

(b) Evolution of eukaryotes from an ancestor of modern planctomycetes, which are bacteria that have cell walls and membrane-bound compartments, one of which contains genetic material mixed with DNA- and RNA-processing proteins (Fig. 11.1).

(c) A separate (now extinct) cell type equipped with a cytoskeleton (the chronocyte) evolved first and engulfed archaea and bacteria to generate the nucleus and the metabolic machinery of the eukaryotic cell.

(d) Similarities between nuclei and viruses, particularly that the latter are essentially packets of DNA surrounded by a protein coat, and often by a membrane, have given rise to the idea that the membrane-bound nucleus and other eukaryotic features originated from the primeval infection of a prokaryote by a virus.

Some of these ideas strongly imply that the nucleus could date back to the LUCA, from which eukaryotes, bacteria and archaea eventually diverged. If this is the case, some features of LUCA, such as the nucleus, were retained in eukaryotes but lost to various degrees in most archaea and bacteria. For my current argument I find it interesting that Penny & Poole (1999) dismiss fusion of a bacterium and an archaean on the grounds that it does not explain the origin of the nuclear membrane 'which is assembled and disassembled during cell division, quite unlike organellar membranes'. As I have already pointed out, this criticism cannot apply to Kingdom Fungi. Characteristically, nuclear divisions in fungi take place within the parental nuclear membrane. Consequently, by whatever route the eukaryotic nucleus arose, its most primitive expression survives in present-day fungi. Perhaps, then, this is the first hint that present-day fungi are the survivors of the most primitive eukaryotes.

The most complete reworking of the tree of life is that recently published by Tom Cavalier-Smith (Cavalier-Smith, 2006, 2010a, 2010b). Cavalier-Smith's approach is to integrate palaeontology with comparative study of present-day organisms, emphasising key steps in molecular and cellular evolution. Cavalier-Smith (2010a) identifies five successive kinds of cell:

(a) The first cells were negibacteria, with cells bounded by two acyl ester phospholipid membranes, divided into the primitive anaerobic Eobacteria without lipopolysaccharide in the outer membrane and more advanced Glycobacteria with lipolysaccharide (e.g. oxygenic Cyanobacteria and Proteobacteria);

(b) unibacteria, with one bounding and no internal membranes, divided into desiccation-resistant posibacteria, ancestors of eukaryotes, and

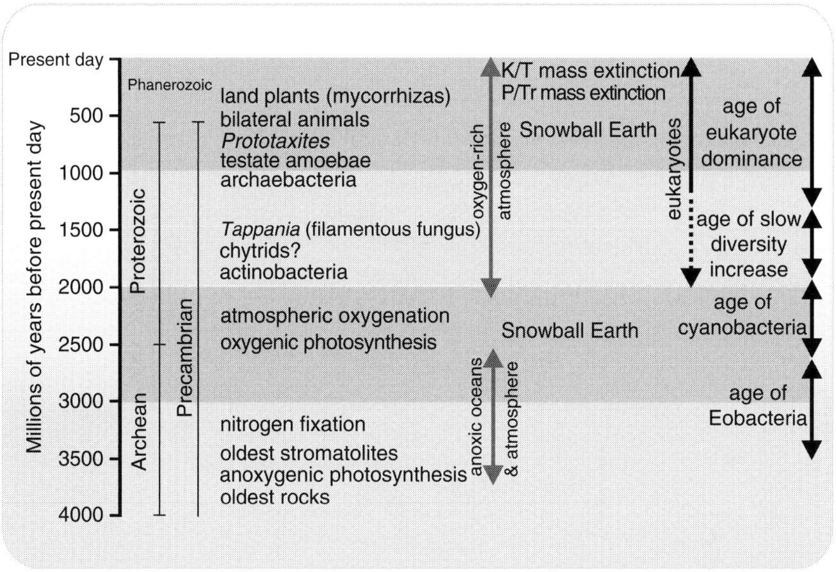

Fig. 11.2 A geological timescale covering from the time of the oldest rocks (3.8 billion years ago) to the present highlighting major geological and evolutionary events and features mentioned in the text, including my interpretation of Cavalier-Smith's four ages of life at extreme right. Note that Cavalier-Smith's age of eukaryotes starts 850–800 million years ago, but as I interpret *Tappania* fossils to be fully differentiated sclerotia of filamentous fungi I place the origin of stem (chytrid) eukaryotes between 2.0 and 1.5 billion years ago. (Modified and redrawn from Cavalier-Smith, 2010a.)

archaebacteria as the youngest bacterial phylum and a sister group (not an ancestor) of eukaryotes;

(c) eukaryotes with endomembranes and mitochondria, (eukaryotes plus archaebacteria make up the neomura);
(d) plants with chloroplasts;
(e) chromists with plastids inside the rough endoplasmic reticulum.

These types of cell are placed into four ages of life as follows (Cavalier-Smith, 2010a; see Figs. 11.2 and 11.3):

(a) The age of Eobacteria, an anaerobic phase in which photosynthetic non-sulphur bacteria (and before them extinct stem negibacteria) were the major primary producers. Major consumers with surviving descendants were heterotrophic chlorobacteria, and perhaps others that preceded the origin of photosystem II. Exclusively anaerobic life probably persisted from about 3.5 billion years ago to just under 2.5 billion years ago (the best date for the origin of photosystem II and start of oxygenic photosynthesis).

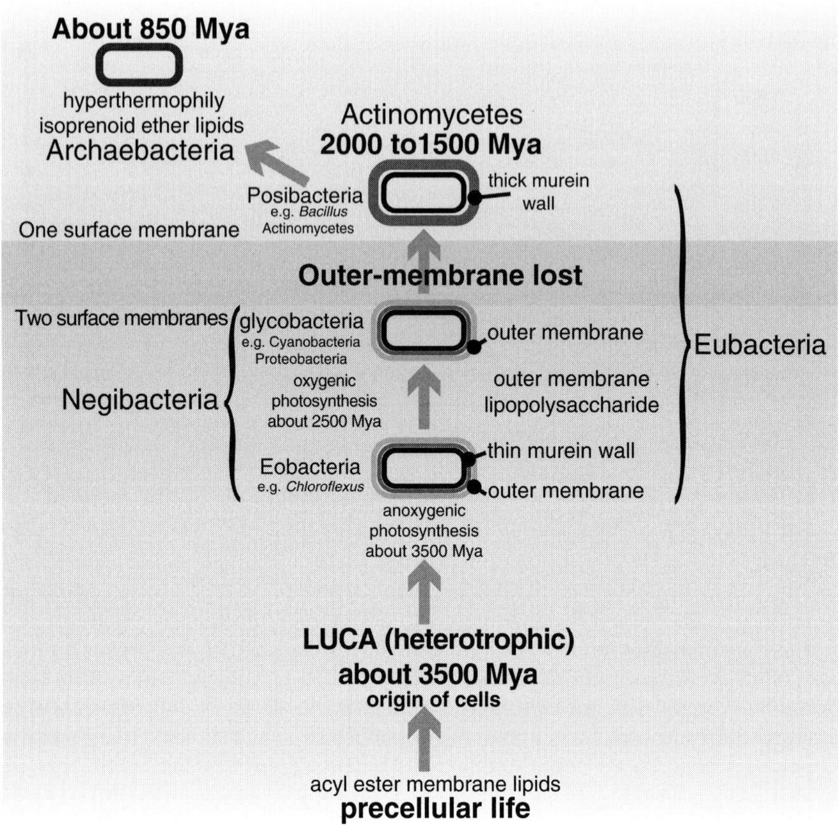

Fig. 11.3 The tree of life. This diagram is based on Cavalier-Smith's tree of life (2010a; his fig. 6) which emphasises major evolutionary changes in membrane topology and chemistry, except that the most ancient bacteria are shown here to be heterotrophic descendants of LUCA (the last universal common ancestor).

(b) The age of cyanobacteria (about 2.5–1.5 billion years ago) during which cyanobacteria were the major primary producers (and are now the dominant morphological fossils). Convincing fossils of various cyanobacteria have been dated to the later part of this period, including complex filamentous forms, some with heterocysts (= nitrogen fixation?). Extensive anaerobic habitats probably remained, especially in the deep ocean. The origin of eubacterial flagella was a major innovation during this age (enabling planktonic existence after release from the ancestral biofilm), and substantial metabolic diversification of chemotrophic and heterotrophic negibacteria.

(c) The age of slow diversity increase (1.5–0.85 billion years ago) features increasing morphological complexity and colonisation of continental surfaces by both Cyanobacteria and, following loss of the outer membrane, Posibacteria and the actinomycete Actinobacteria; the latter displaying the greatest morphological complexity. Some of the largest microfossils from this part of the middle Proterozoic have been attributed to eukaryotic algae, filamentous fungi or stem eukaryotes of undefined affinity, but Cavalier-Smith is sceptical of all such fossil identifications in this period.
(d) The age of eukaryotes and obvious macroorganisms (850–800 million years ago to the present according to Cavalier-Smith, but in my view from between 2.0 and 1.5 billion years ago to the present). Cavalier-Smith (2006) argues that eukaryotes derived from an actinobacterial ancestor on the grounds (among others) that current Actinobacteria are the only eubacteria having phosphatidylinositol, which is one of the most important eukaryote phospholipids, required for eukaryote specific cell signalling. 'Thus, eukaryote membrane lipids probably came vertically from an actinobacterial ancestor, archaebacterial lipids originating in their [last common ancestor] after it diverged from eukaryotes.' A further aspect of this argument is that shortly after they diverged from eukaryotes archaebacteria colonised hot, acid environments by evolving the ancestrally hyperthermophilic archaebacteria and later, one archaebacterial lineage evolved biological methanogenesis (Cavalier-Smith, 2006, pp. 977–978).

This last description (paragraph (d)) encapsulates the revolutionary differences between the Cavalier-Smith model and the 'standard' three-domain model based on Woese *et al.* (1990).

(a) The standard model perceives the archaebacteria as an ancient (over 3.5 billion years old) group of prokaryotes which was the ancestor of eukaryotes.
(b) The Cavalier-Smith model sees the Archaebacteria as sisters to eukaryotes, rather than their ancestors, and, moreover, as a group that first diverged only 800 million years ago.

This difference also has major implications for the last universal common ancestor (LUCA). The standard three-domain model gives credence to the belief that LUCA emerged from the iron-sulphur world of deep, hot hydrothermal vents, which specifically means that LUCA was a hyperthermophile. But in the Cavalier-Smith model this cannot be true because hyperthermophiles are assumed to have appeared for the first time less than 800 million years ago; so this leaves open the possibility (which I believe to be true) that LUCA was a mesophile that arose in a temperate environment.

Generally speaking I find the Cavalier-Smith model much more convincing because it is based on integration of such a broad range of data. So I accept Cavalier-Smith's narrative from the first appearance of living cells about 3.5 billion years ago to the emergence of eukaryotes from an actinobacterial ancestor about 1 billion years ago (both dates give-or-take the odd 100 million years). I part company with his version of the origin of eukaryotes for reasons that will be clear from the following quotations:

> ... radical innovations in cell structure that made eukaryotes were tied up with the origin of predation on other cells by engulfing them by phagocytosis, an ancestral property for protozoa and animals. By contrast, no bacteria can eat other cells by engulfment, though several groups of bacteria became predators by evolving enzymes to digest prey externally, just as do some fungi and carnivorous plants. (Cavalier-Smith, 2010a, p. 114)
>
> Protozoa became the major predators on bacteria in water and wet earth; typically brownish photophagotrophic and photosynthetic chromists conquered the oceans; a green alga became a land plant 400 Myr ago, its descendants coating the continents where not too dry or cold with a green veneer providing homes and food for descendants of mobile animals (bilateria) that evolved 530 Myr ago via Cnidaria from marine sponges that fed on bacteria, like their choanoflagellate protozoan ancestors ... One choanoflagellate created sponges by evolving epithelia and connective tissue to allow more extensive filter feeding, and anisogamous sex to allow nonfeeding ciliated larvae to grow large before settling onto rocks to feed. A distant choanozoan relative encased its filopodia in chitinous walls to evolve fungi that colonized soil as saprotrophs on dead plant material and symbionts and parasites of land plants. (Cavalier-Smith, 2010a, p. 125)
>
> Thus, when fungi and oomycetes evolved, ancestral protozoan phagotrophy was lost through the origins of their cell walls ... (Cavalier-Smith, 2010a, p. 119)

There are many other paragraphs in similar vein, especially in Cavalier-Smith (2010b), but those quoted are sufficient for my purpose. I think this interpretation is wrong because it is:

(a) animal centric, my antipathy for which I have already clearly indicated; but this interpretation seems to go even further by quite unjustifiably equating the eukaryotic grade of organisation with animal phagocytosis;
(b) dismissive of fungi, treating them essentially as failed animals that arrived on the world scene after the plants had invaded the land;
(c) totally dependent on 'the origin of predation on other cells by engulfing them by phagocytosis, an ancestral property for protozoa and animals' though this extreme position is taken without suggesting what selective

advantage there might be in the essential intermediate evolutionary steps towards phagocytosis.

Any advantage there might be in recognising a neighbouring cell as 'prey' cannot be realised until the (quite complicated) phagocytosis process is fully assembled. The advantage of predation does not become apparent until:

(a) the membrane of a cell otherwise enclosed by a wall can be exposed to the environment without consequential osmotic stress;
(b) exposed membrane can detect 'prey' and be shaped to engulf it and then be withdrawn (as a food vacuole) into the predator's cytoplasm; and
(c) the food vacuole is equipped with enzymes to digest the prey (without digesting the predator) and later with mechanisms to distribute the extracted nutrients and dispose of waste products.

All of this requires water management, extreme membrane management of endocytosis and exocytosis, and complete cytoskeletal management of enzyme, vesicle and vacuole movement and distribution. Although the selective advantage of such a process is self-evident now, I can't see how any advantage can be realised by some distant animal-ancestor that is just embarking on acquiring these characters. But I think I can see how a fungus might do it, and Martin *et al.* (2003) saw at least part of the way:

> The view that osmotrophy had to precede phagotrophy in eukaryotic evolution is compelling because without importers, food vacuoles are useless ... all fungi are osmotrophs ... (Martin *et al.*, 2003, p. 199)

I believe that eukaryotes emerged from primitive biofilms about 2 billion years ago. Most of my discussion of this appears in the next chapter, but to provide a link with the biofilm discussion in Chapter 10 I want to emphasise that fungi can participate in biofilms in the present day. Initially, of course, the stem eukaryotes would have been single-celled organisms so it is important to stress that baker's yeast, *Saccharomyces cerevisiae*, can initiate biofilm formation. Growth in low-glucose medium causes the yeast cells to adhere to plastic surfaces, and to form 'mats' on dilute agar medium where yeast-form cells adhere together. Both formation of mats and attachment to plastic require a particular family of fungal cell surface glycoproteins for adherence (Reynolds & Fink, 2001). Filamentous fungi also form biofilms in the present day (Harding *et al.*, 2009) and the more primitive zygomycetes, including *Rhizopus* and *Rhizomucor*, produce an extensive extracellular matrix to aid adherence to surfaces (Singh, Shivaprakash & Chakrabarti, 2011).

It seems likely, therefore, that as soon as they emerged fungi would have formed part of the primitive biofilm community. It is consequently possible to envisage that the biofilms of about 2 billion years ago were able to confine together the prokaryotes that needed to collaborate to form the first unicellular stem eukaryotes. In essence, then, the extensive biofilms of 2 billion years ago served to foster the emergence of eukaryotes, just by doing what biofilms do (Chapter 10). Further, the biofilms would have also fostered the development of filamentous fungi by providing the appropriate selection pressure for the filamentous growth form to escape from and destructively exploit and dominate the biofilm matrix. An important ability of hyphal growth is that apically growing filamentous hyphae can explore and exploit the biofilm, digesting the adhesives and other polymers that make up the biofilm matrix, even escaping from the biofilm under their own volition. Even more importantly, filamentous hyphae can parasitise the photosynthetic microbes of the biofilm community to recruit photobionts into primitive lichen-like arrangements, which then have the terrestrial surface of the Earth at their disposal.

TWELVE

RISE OF THE FUNGI

Although fungal hyphae have few unique morphological features and most fungal structures are poor candidates for preservation over long periods of time as fossils, a respectable fossil record for fungi has been assembled in recent years. By far the most impressive fungi of the Ordovician/Devonian Period are specimens of the fossil genus *Prototaxites*, which were terrestrial organisms found from the mid Ordovician (460 million years ago) to the early Devonian, suggesting that they lasted a period of at least 40 million years (Boyce *et al.*, 2007; Hueber, 2001). These fossils are among the 'nematophyte phytodebris' that constitutes the earliest evidence for terrestrial organisms. This 'nematophyte phytodebris' nomenclature was assigned in the middle of the nineteenth century and has no relevance to present-day understanding of taxonomy (that is, it does *not* necessarily indicate that the stuff was of plant origin); though it does indicate that confusion over the identification of the material is over 150 years old (see discussion in Hueber, 2001 and Taylor *et al.*, 2010). *Prototaxites* specimens are generally large: over a metre wide (Wellman & Gray, 2000) and up to 8 m tall (Hueber, 2001) (Figs. 12.1 and 12.2; illustrated in colour in Moore *et al.*, 2011, pp. 33–34). *Prototaxites* was also so common that it was a major component of these early terrestrial ecosystems, both in terms of abundance and diversity. Some of the earliest examples found were tree-like trunks constructed of interwoven tubes < 50 µm in diameter (concentrically arranged in transverse sections), and the fossils were interpreted to be small coniferous trees, though we now know that

Fig. 12.1 Two views of Lower Devonian *Prototaxites* compression fossils, approximately 2 m tall, *in situ* in the Bordeaux Quarry, near Cross Point on the Restigouche River in the Gaspésie region of eastern Quebec, Canada. What you are looking at here is a stream bed that is turned more or less on its side (that is, at right angles to its original position), so that you are now looking down on it. You can see fossilised impressions of at least three large *Prototaxites* specimens that must have formed something like a log jam in the stream. Dr Francis Hueber, who first made the suggestion that *Prototaxites* fossils are fungal in origin (Hueber, 2001), is posing alongside as a scale marker. The two photographs are identical, but the *Prototaxites* specimens are outlined in the image on the right. (Image kindly supplied by Dr Carol Hotton, courtesy of the Smithsonian Institution; the image appears as fig. 1A in Boyce *et al.*, 2007.)

environments at the time *Prototaxites* was fossilised did not (yet) include large vascular plants.

Prototaxites was by far the largest organism present in these ancient habitats – environments that did not include vascular plants, but were still dependent on the more ancient primary producers: cyanobacteria (blue-green algae), eukaryotic algae, lichens, and mosses, liverworts and their relatives (bryophytes). Isotope ratio mass spectrometry of individual

Fig. 12.2 An artist's impression of the landscape of the Devonian period of some 400 million years ago, dominated by specimens of *Prototaxites* up to 9 metres tall. This is a monochrome rendition of a painting by Geoffrey Kibby that appeared as a rear cover image on the magazine *Field Mycology* in April 2008. In the landscape portrayed in the painting the fungus *Prototaxites* dominates as the largest terrestrial organism to have lived up to this point in time. Although vascular plants had emerged by this time, these landscapes were still dependent on the more ancient primary producers: cyanobacteria (blue-green algae), eukaryotic algae, lichens and mosses, liverworts and their bryophyte relatives. What you are seeing here is the physical expression of the dominance of fungi in the Earth's biosphere. This physical dominance of *Prototaxites* lasted at least 40 million years (about 20 times longer than the genus *Homo* has so far existed on Earth). (Image kindly supplied by Geoffrey Kibby, senior editor of *Field Mycology*.)

Prototaxites fossils provides the most compelling evidence that *Prototaxites* was a fungus (Boyce *et al.*, 2007; Hobbie & Boyce, 2010); I personally consider this evidence conclusive. These analyses show that carbon isotope ratios (^{12}C:^{13}C) of individual *Prototaxites* fossils varied too much for them to be photosynthetic primary producers of any sort (Boyce *et al.*, 2007), indicating that *Prototaxites* was a heterotroph (saprotroph) that digested isotopically heterogeneous substrates; it was a consumer and recycler. Hobbie & Boyce (2010) demonstrated a similar large range of carbon isotope values among fungi, particularly saprotrophic fungi, of a present-day environment resembling the Early Silurian and Devonian landscapes where *Prototaxites* occurred. Hueber's (2001) critical examination of the microscopic anatomy of *Prototaxites* found similarities with the trimitic system of hyphae evident in present-day basidiomycetes; he states:

> *Prototaxites* is nomenclaturally valid ... This report has a triple purpose: (1) to name, as neotype, a recognizable specimen [of *Prototaxites*] collected by Dawson for which the locality and stratigraphic data are known, (2) to redescribe the genus as structurally composed of three interactive forms of hyphae, i.e. large thin-walled, septate, branching, generative hyphae; large thick-walled, non-septate, skeletal hyphae; and small thin-walled, septate, branching, binding hyphae, which combine to form a gigantic, phototropic [more likely gravitropic in my view], amphigenous [i.e. a hymenial hyphal layer of present-day Ascomycota and Basidiomycota that extends over the entire surface of the spore-producing body], perennial sporophore with saprobic nutrition, and (3) to classify it in the Kingdom Fungi. (Hueber, 2001, abstract [with comments from me in square brackets])

Taking all this evidence together the conclusion is inescapable to me that these enormous fossils, which were the largest land organisms to have lived up to their point in time, were actually giant terrestrial saprotrophic fungi, with affinities (dolipore septa, clamp connections and sterigmata) to present-day Basidiomycota.

Other ancient fungal fossils are found in the exquisitely preserved Devonian Rhynie Chert of Aberdeenshire in the north of Scotland, which is over 410 million years old and contains fossils of primitive plants (with water-conducting cells and sporangia, but no true leaves), along with arthropods, lichens, algae and fungi. Several fungi, indeed, representing close to a complete range of the lifestyles and developmental patterns seen in fungi of the present day (Fig. 12.3) have been found in this sedimentary deposit (Taylor, Hass & Kerp, 1997; Taylor *et al.*, 2004; Taylor, Krings & Kerp, 2006).

The fungal lifestyles represented in the Rhynie Chert specimens shown in Fig. 12.3 range through the following:

(a) a member of the Oomycota, which is a phylum of filamentous protists known as water moulds or downy mildews, which are not fungi but primitive fungus-like organisms. Many of the Oomycota are important plant pathogens in the present day (such as *Phytophthora* species). Several of the Rhynie Chert specimens show an antheridium contacting the oogonium, clearly demonstrating the operation of a fully differentiated sexual reproduction process including differently differentiated male and female gametes and all that goes with them, including sexual hormones, hormone receptors and male to female cell targeting.

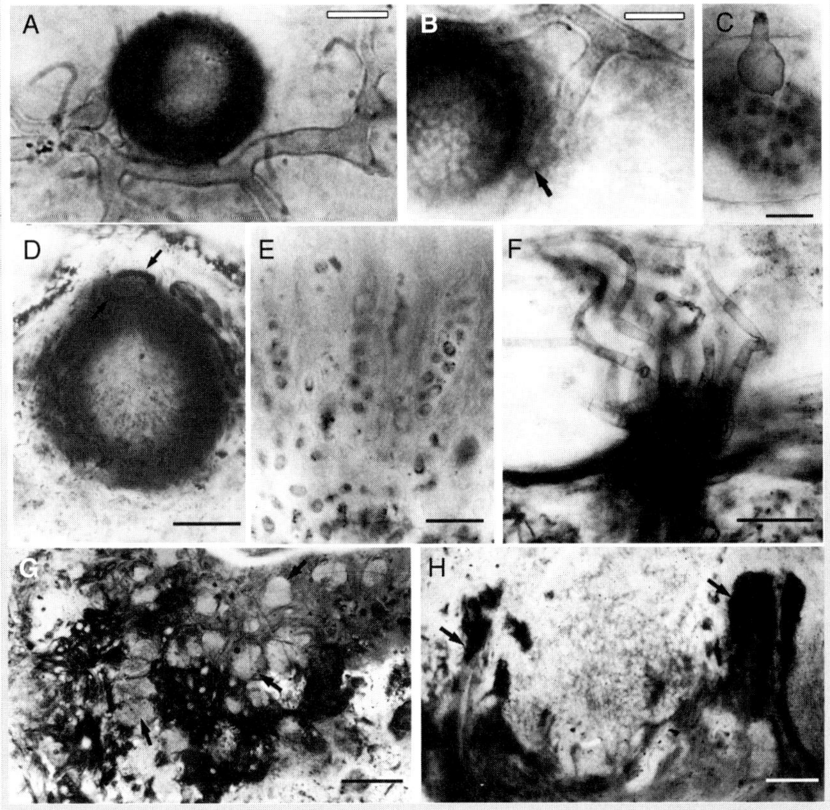

Fig. 12.3 Ancient fungal fossils found in the Devonian Rhynie Chert of Aberdeenshire in the north of Scotland (410 million years old). **A** and **B**, *Hassiella monospora*, a member of the Peronosporomycetes (Oomycota) associated with plant debris. **A**, branching hyphae and thick-walled oogonium/oosporangium; scale bar = 10 μm. **B**, mature oogonium (or immature oosporangium) with hyphal branch differentiated as antheridium (at arrow) in contact with the spore wall surface revealing a fully differentiated sexual reproductive system; scale bar = 10 μm. (Taken from Taylor, Krings & Kerp, 2006, with permission from Elsevier © 2006.) **C**, chytrid zoosporangium with neck extending through the wall of a plant cell; scale bar = 15 μm. **D**, longitudinal section of perithecium showing central cavity containing asci. Note guard cells (arrows) surrounding ostiole; scale bar = 100 μm. **E**, several asci containing ascospores from Rhynie Chert ascomycete; scale bar = 15 μm. **F**, tuft of conidiophores erupting from epidermis of *Asteroxylon* stem; scale bar = 25 μm. **G**, section of cyanolichen thallus showing depressions (lighter areas at arrows) that

(b) Another specimen shows fully differentiated chytrid zoosporangia and carries the inference that at the time the sediments were laid down the complete cell biology of the chytrids (thallus, rhizoids, free cell formation, motile zoospores, etc.) was fully established. Also among the Rhynie Chert specimens are examples of chytrids parasitising other fungi, showing, again, that 400 million years ago the fungal lifestyle was so firmly established that fungi were parasitising other fungi.
(c) Higher fungi are also represented in the Rhynie Chert, with particularly fine ascomycete specimens (Fig. 12.3) showing ascospore development typical of the present day, and differentiated conidiophores emerging through a plant epidermis; suggesting a fully adapted plant pathogen at an extremely early stage in plant evolution. However, although the associated plants are at an early stage, the ascomycete fungal fossils appear to be highly evolved. The hyphae are regularly septate; the specimens include perithecia (ascomycete fruit bodies) apparently identical to those of the present day, from which we can infer a fully evolved fungal developmental biology able to produce extreme hyphal differentiation, cell signalling, cell sorting, pattern formation and formation of tissues with different functions, all of which are typical of the present day.
(d) Figure 12.3 also shows sections of a cyanolichen thallus, which is again indicative of a highly evolved fungal developmental process at this very ancient time. Evidently, fungi were able to entrap cyanobacteria, create differentiated fungal thalli and invade the land 400 to 500 million years ago.

This ancient fungal extravaganza wasn't just happening in Scotland. Glomeromycotan fossils have been found in mid Ordovician rocks of Wisconsin which are 460 million years old (Fig. 12.4). The fossilised material consisted of entangled, occasionally branching, non-septate hyphae together with globose spores. The age of these fossil glomeromycotan fungi indicates that such fungi were present before the first vascular plants arose, when the land flora consisted of bryophytes, lichens and cyanobacteria. Today, the Glomeromycota form the arbuscular mycorrhizal symbiosis, which is ubiquitous in modern vascular plants and has also been reported in modern

Caption for Fig. 12.3. continued

contain cyanobacteria surrounded by more opaque zone that represents the fungal partner; scale bar = 750 μm. **H**, section of lichen at right angles to thallus in **G**, showing depression and opaque walls (arrows) formed of fungal hyphae; scale bar = 150 μm. (Images C to H taken from Taylor *et al.*, 2004, and reproduced by permission of The Royal Society of Edinburgh from *Transactions of the Royal Society of Edinburgh: Earth Sciences*, volume 94 (2004, for 2003), pp. 457–473.)

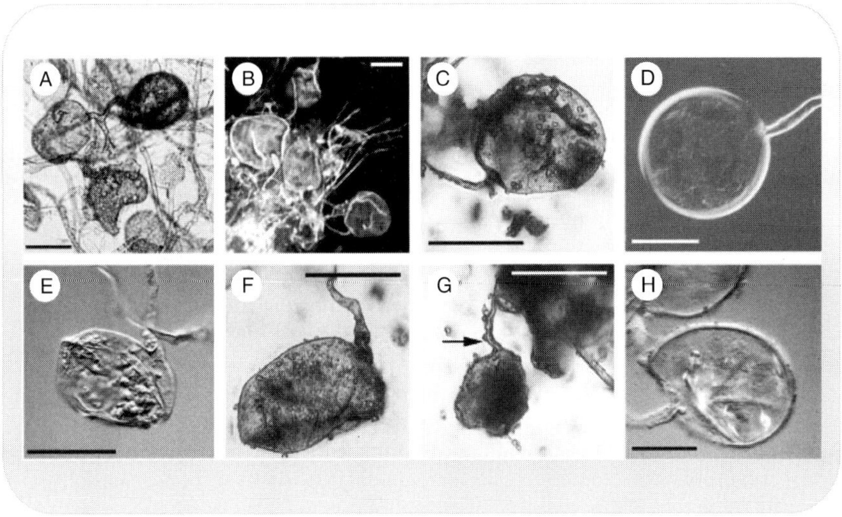

Fig. 12.4 Comparison of fossil and current glomalean fungi. Fossil specimens (**A** to **C** and **E** to **G**) found in the Guttenberg Formation, mid Ordovician dolomite of Wisconsin, which was deposited between 460 and 455 million years ago, compared with spores formed by present-day glomalean fungi (**D** and **H**). **A** and **B**, overviews of the fossilised material. **C**, **E**, **F** and **G**, fossil spore details. **C**, detail of **B**. **D**, a spore of present-day *Glomus* sp. with layered wall structure. In **G**, the arrow shows walls of a subtending hypha in connection with the spore wall. **H**, a spore of present-day *Glomus leptotichum*. All scale bars are 50 µm. (Image and legend adapted from Redecker, Kodner & Graham, 2000. Reprinted with permission from AAAS.)

hepatics and hornworts. It is reasonable to suppose that arbuscular mycorrhizas played an important role in the success of early terrestrial plants (Blackwell, 2000; Redecker, Kodner & Graham, 2000); in other words, the mycorrhizal fungi enabled the plants to invade the land.

A recent textbook continues this story in this way:

> Other convincing fossils are more recent. Microfossils of hyphae with clamp connections are known from the Pennsylvanian/Carboniferous (300 million years ago . . .). However, the only convincing mushroom fossils found so far are preserved in amber dating from the Cretaceous (about 90 million years ago). These 'mushrooms in amber' (Hibbett, Grimaldi & Donoghue, 1995) are particularly interesting because they bear a strong resemblance to the existing genera *Marasmius* and *Marasmiellus*, which are quite common in modern woodlands; yet when the fossils were preserved the dinosaurs still ruled the Earth. In other words: the mushrooms you see when you trek through the forest are almost identical to those seen by dinosaurs in their forests. Although, of course, the forest plants are very different [the animals are somewhat less alarming, too].

> Amber dated to the Eocene (54–34 million years ago ...) has been found that contains the remains of several filamentous mould fungi. Among the finds are sooty moulds in European amber dating back to 22–54 million years ago. Present day sooty moulds are a mixed group of saprotrophs, usually with dark coloured hyphae, which produce colonies superficially on living plants (as harmless epiphytes). Most present day sooty moulds use arthropod excretions for nutrition and live closely associated with aphids, scale insects and other producers of honeydew. All the fossils are composed of darkly coloured hyphae with features identical to the present day genus *Metacapnodium*, suggesting that *Metacapnodium* hyphae have remained unchanged for tens of millions of years (Rikkinen *et al.*, 2003). Possibly the most impressive fossil is a piece of amber from the Baltic region that contains an inclusion of a springtail (a collembolan arthropod) which is overgrown by an *Aspergillus* species (Dörfelt & Schmidt, 2005). The surface of the springtail is densely covered in places by excellently preserved hyphae and conidiophores. Numerous sporulating conidiophores can be seen easily, and conidial heads with radial chains of conidia are clearly visible. As well as superficial hyphae at the cuticle, the springtail is loosely penetrated by branched substrate hyphae, so the authors suggest that the fungus may be parasitic and describe it as a new species, *Aspergillus collembolorum* (Dörfelt & Schmidt, 2005). (Moore *et al.*, 2011, pp. 31–32)

Thus, convincing fossil evidence shows that fungi were important, even dominant, members of terrestrial ecosystems around the globe at least 500 million years ago. Well-developed filamentous fungi must have first appeared a long time before that, however. How long would it take the ancestors of the Rhynie Chert water moulds, chytrids, Glomeromycota, lichens and Ascomycota to evolve the capability to form structures microscopically indistinguishable from those of the present day? How long would it take the ancestors of *Prototaxites* to evolve an 8-metre tall club-fungus, and become distributed worldwide? Guessing could push 'well-developed filamentous fungi' back in time to about 700–800 million years ago. But there are older fossils than that, even though others may dispute them.

Butterfield (2005) assigned fossils (illustrated in Fig. 12.5) from formations in northwestern Canada, the deposition of which has been dated to between 800 and 900 million years ago, to the form-genus *Tappania*; describing the organism as:

> ... an actively growing, benthic, multicellular organism capable of substantial differentiation. Most notably, its septate, branching, filamentous processes were capable of secondary fusion, a synapomorphy of [trait shared by] the 'higher fungi' [of today]. Combined with phylogenetic, taphonomic and functional morphologic evidence, such 'hyphal fusion'

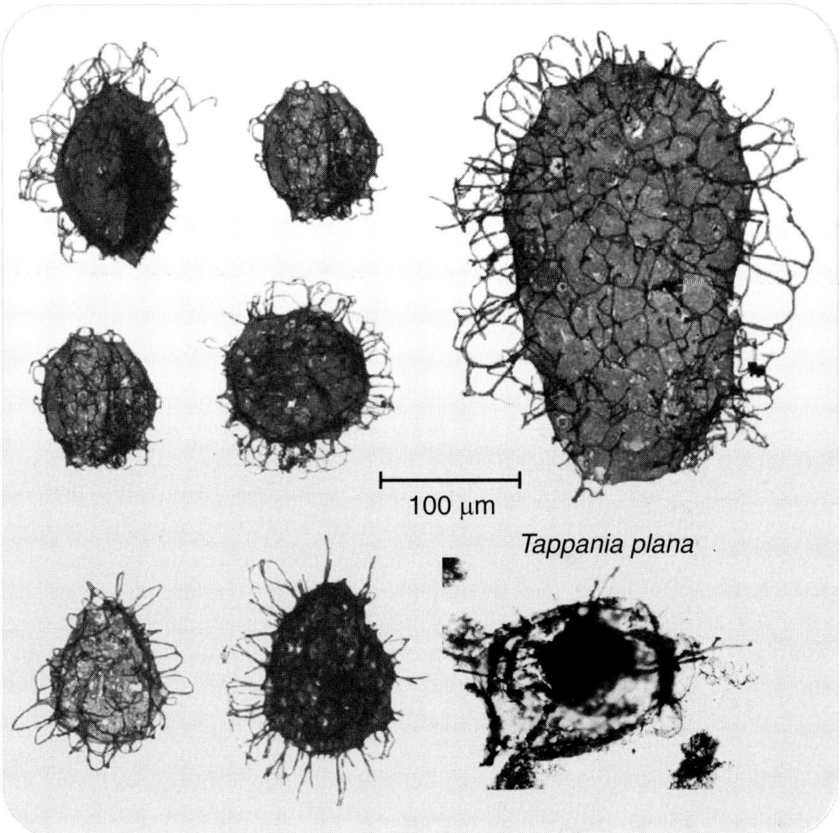

Fig. 12.5 *Tappania* sp. fossils. All except the specimen at bottom right are from the Wynniatt Formation on Victoria Island, northwestern Canada. Specimens are generally described as 'vesicles' with 'processes'; some are described as 'now flattened'. These specimens are between 800 and 900 million years old. (Images republished with permission of Paleontological Society, Inc., from Butterfield, 2005; permission conveyed through Copyright Clearance Center, Inc.) The image shown at bottom right is *Tappania plana*, from shales of the early Mesoproterozoic Roper Group in northern Australia. This specimen is 1492 to 1429 million years old. (Image taken from Javaux, Knoll & Walter, 2001, and reprinted by permission from Macmillan Publishers Ltd: Nature © 2001.) All images have been adjusted to the same scale.

> identifies *Tappania* reliably, if not conclusively, as a fungus, probably a sister group to the 'higher fungi' [Dikarya], but more derived than the zygomycetes. (Butterfield, 2005, abstract)

The form-genus fossil *Tappania* is widespread, having been found in ancient shoreline carbonaceous shale deposits in Australia, Canada and China.

Specimens fossilised nearly 1.5 billion years ago in shales in northern Australia have been described as follows:

> *Tappania* populations consist of irregularly spheroidal organic vesicles up to 160 µm in diameter ... distinguished by bulbous protrusions and from zero to twenty hollow, cylindrical processes ... The processes have closed, slightly expanded terminations and may branch dichotomously ... processes are distributed irregularly and asymmetrically on the vesicle surface ... the irregular number and length, asymmetric distribution, and branching of processes in *Tappania* suggest an actively growing cell or germinating cyst. The bulbous protrusions in some specimens further suggest vegetative reproduction through budding ... (Javaux, Knoll & Walter, 2001)

The asymmetric branching of processes and bulbous protrusions are interpreted by Javaux *et al.* (2001) as representing dynamic cell remodelling of a sort which is only made possible by the cytoskeleton and signalling pathways of eukaryotes. Javaux *et al.* (2001) go no further than to state that the systematic relationships of *Tappania* are uncertain, but its distinctive morphology indicates that 'the cytoskeletal architecture and regulatory networks that characterize living [eukaryote] protists' were in place in organisms fossilised 1.5 billion years ago (see Fig. 12.5). However, Butterfield (2005) discusses these and other putative pre-Devonian fungi and concludes that 'there is a case to be made for an extended and relatively diverse record of Proterozoic fungi.' Cavalier-Smith (2006, pp. 983–984) agrees with Butterfield's (2005) identification of *Tappania* as sporangial entities broken from a branching trophic hyphal network, but does not agree that these fossils are probably fungi. He suggests they could instead be actinobacterial pseudosporangia; I do not find this very convincing.

The large spheroidal microfossils shown in these *Tappania* papers (several samples shown in Fig. 12.5) are usually described as 'vesicles'. Butterfield's (2005) specimens, after being dissolved into slurry with 30% hydrofluoric acid and filtered through a 62 µm mesh sieve, are described as follows:

> The fossils described here constitute a highly variable, bimodal continuum of forms. Those of the principal mode are based on a central vesicle bearing a variable number of irregularly distributed processes and occasional larger-scale outgrowths. The central vesicle ranges from spheroidal to elongate, and from 30 µm ... to over 400 µm ... in transverse dimension ... Processes are typically heteromorphic and range from 0.3 µm ... to > 4 µm ... in diameter. In some instances, simple cylindrical processes may be distributed relatively uniformly over the vesicle surface ...; in others, they occur as isolated knoblike buds ... or elongate filamentous extensions ... In most cases, however, the processes are further distinguished by distal branching ... and a capacity to form closed loops through secondary fusion. This fusion appears to be

> relatively indiscriminate and gives rise to a wide range of expression: occasionally the processes return directly to the vesicle to form simple loops ...; in other cases they have fused either with themselves ... or, more commonly, with other processes ..., resulting in a distally interconnected network ... Multiple layers of process networks are also developed, sometimes to the extent of obscuring the central vesicle ... Such variability, combined with a recurrence of unfused buds – on both the vesicle ... and processes ... attests to the actively growing habit of these structures. (Butterfield, 2005, p. 167)

This is quoted in detail because I have spent most of my research life cultivating a basidiomycete fungus (recently renamed *Coprinopsis cinerea* but called *Coprinus lagopus* or *Coprinus cinereus* for most of the time I was cultivating it) which, in common with many other present-day ascomycete and basidiomycete soil fungi, produces abundant **sclerotia** in and on mycelial cultures. A **sclerotium** is a compact mass of fungal hyphae woven into a vegetative food-storage body (Fig. 12.6). Its purpose is simply to survive hard times; in most cases this means it is an overwintering structure, but sclerotia can be produced throughout the year and can survive hot, dry summers, too. To aid their survival the outermost layer of hyphae are dark coloured with a melanin pigment and the whole structure is hardened by active exudation of water. The sclerotium detaches from its parental mycelium when mature and can be distributed by wind and rain. When conditions improve the structure serves as inoculum for a fresh mycelium when new hyphae grow out of the sclerotium using the stored food sources it contains to support the initial new growth. This simple description is deceptive. So deceptively simple that the developmental significance of sclerotia is rarely appreciated. Let's examine the description above to extract that significance explicitly:

(a) **Compact mass of fungal hyphae**: fungal hyphae are exploratory, they grow away from each other; when a hypha forms a branch, the branch grows away from the parent hypha; technically it's called negative autotropism, a tropism being growth in a specific direction, and a negative tropism is growth away from something, and as auto means 'self', negative autotropism means that the hyphae of a mycelium naturally grow away from the rest of the mycelium. Consequently, the only way a mycelium can make a compact mass of hyphae is to reverse this tropism, to make branches grow towards their parent hypha – from which we can infer that a fungus that makes sclerotia is sufficiently advanced to have:

(i) an autotropic mechanism (something that detects the presence of sister hyphae and something that instructs the Spitzenkörper (see Chapter 2) to react to that presence);

(ii) a regulatory system able to change from the default negative autotropism to the cell-targeting positive autotropism;
(iii) possibly, a cell-to-cell adhesion system to help stick the hyphae into a compact ball;
(iv) sufficiently refined ability to localise differentiation along the hypha to ensure all this only happens at specific places;
(v) signalling systems that react to environmental and intracellular cues to ensure all this only happens where and when the mycelium has the ability to complete the whole process.

(b) **Food-storage body**: sclerotia store carbohydrate polymers; initially glycogen is stored in globose thin walled cells, but during the maturation process the glycogen is remobilised and other cells convert the released sugars into a secondary wall made of thick glucan fibrils. Evidently, individual cells, although they are just compartments along the length of the hyphae that are woven into the sclerotium, are differentiated into different types of cells where sophisticated metabolism can convert nutrients to storage compounds, the final one of which is a cell wall component. Being a food-storage body implies the accumulation of more nutrients in the sclerotium than can be used to construct the body of the sclerotium. Since the hyphae within the developing sclerotium cannot extract these nutrients from their surroundings for themselves, this further implies a translocation network that can bring nutrients to the sclerotium from elsewhere in the mycelium. This further implies that there is a differentiated class of mycelial hyphae that are targeted on, and dedicated to, the developing sclerotium to provide it with nutrients rather than contribute directly to its structure (Fig. 12.6) – from which our inferences are:
(i) there is an efficient long-distance translocation of nutrients along differentiated hyphae;
(ii) the hyphae that make up the central zones of the sclerotium differentiate into at least two types of cell that have a metabolism able to accumulate polysaccharide in the form of either glycogen granules or glucan fibrils in a secondary cell wall;
(iii) outermost layer(s) of hyphae are dark coloured: this is another feature that implies a differentiation of hyphal cells; in this case it also implies a regulatory system that restricts the differentiation specifically to the outermost layers of the sclerotium, and in some sclerotia literally to the one, single, outermost layer of cells. The differentiation involves wall thickening, which is again localised preferentially to wall on the outer surface of the sclerotium, and which is pigmented brown to black with melanin – from which we can infer:
(iv) another cell differentiation pathway;

(v) a mechanism that detects and informs cells specifically on the outer surface of the sclerotium;

(vi) a similar mechanism that operates within the cell to direct wall thickening and pigmentation only to the wall(s) on the outer surface of the sclerotium;

(vii) activation of a branch of metabolism to make and deposit the necessary pigment as a secondary layer in the appropriate walls;

(viii) active exudation of water: under normal circumstances water enters fungal hyphae by osmosis (because the cytoplasm contains more solutes than the surrounding fluid) and continues to enter until the containing wall is stressed so that it applies a wall pressure (known as 'turgor') that is equal to the osmotic potential pressure generated by the solute imbalance across the membrane. Droplets of fluid 'exudate' are often seen associated with sclerotia in their final stages of maturation. For water to be exuded like this it must be exported out of the cell against the osmotic flow. This might be a function of the glucan secondary wall in cells accumulating glucan fibrils. The secondary wall is outside the cell membrane, so as the wall increases in volume the cytoplasm is compressed (that is, turgor increases) and water might be squeezed out. Then again, hyphae can translocate water over long distance and water exudation from maturing sclerotia may be done as an adaptation of the nutrient translocation process on which the sclerotium depends for its other nutrients. Although we can't be specific about the process, we can infer from water exudation from the sclerotium that a mechanism operates that is able to export water from specific hyphal cells against the osmotic gradient;

(c) **When conditions improve**: the usual story is that, in nature, sclerotia are made as winter approaches; that is, as temperatures decline and day length decreases. In the laboratory it can also be demonstrated that reduced light and reduced temperature tend to enhance sclerotium production. However, as discussed above, sclerotia are physiologically desiccated so moisture availability may also be involved.

(i) From this we can infer that the sclerotium possesses a means to detect and act upon environmental variables.

(d) **Using the stored food sources**: when conditions do improve a few cells within the sclerotium are activated to start mobilising the stored nutrients and 'germinate'; that is, resume active hyphal growth. Small sclerotia tend to form a new mycelium; larger sclerotia may have sufficient resources to form a structure that makes spores, even a sexual fruiting body like a mushroom.

(i) The inference in this case is that the sclerotium has a mechanism to bring it out of its quiescence; something that enables resumption of normal metabolic activity and normal hyphal growth after the crisis has passed.

From such a simple structure we can draw a considerable range of inferences. A sclerotium is not any old mass of hyphae, then, but a carefully constructed survival machine (Fig. 12.6), from which we can infer a highly evolved metabolism, cell biology and developmental biology.

Let's get back to the sclerotia with which I am most familiar. In Petri dish cultures, *Coprinopsis cinerea* forms sclerotia both in the aerial mycelium that grows above the surface of the agar medium and in the submerged mycelium that grows below the agar surface. Submerged sclerotia were irregularly shaped and about 0.5–1.0 mm diameter. The only structure which differentiated the submerged sclerotium from the submerged mycelium was the sclerotium's outer rind, a single layer of cells with thick, pigmented walls. The central (medulla) region contained the same cell types in the same frequencies as the general submerged mycelium (so something in the submerged mycelium prescribes the boundary of the sclerotium and differentiates the pigmented rind).

In sharp contrast the aerial sclerotia were highly organised structures composed of distinct and compact tissues; though they were polymorphic, meaning that different isolates produced sclerotia with characteristically different internal anatomy, the polymorphisms being genetically determined (Hereward & Moore, 1979). Mature aerial sclerotia were dark brown to black spheroidal structures up to 0.5 mm diameter. An outer layer of dead and moribund hyphae surrounded the main body of the sclerotium which was bilayered with an outer rind and inner medulla. The rind was multilayered in one morphotype, a single cell layer in another morphotype, and consisted of small cells with thick pigmented walls; intercellular spaces were cuticularised (filled with pigmented wall material). The medulla was a closely packed tissue composed predominantly of hyaline thick-walled cells of the same type as were encountered in the submerged mycelium.

In addition to sclerotia, a layer of cells with pigmented thick walls (called brown matting or rind) which differentiated at the air/agar interface was interpreted as an aspect of sclerotial behaviour since it was regularly formed by strains which produced submerged sclerotia and was composed of cells of similar structure to those of the outermost layer of the submerged sclerotium. Aerial sclerotia originated from cells of aerial hyphae (Fig. 12.6A to D). Initially only a single cell was involved. Repeated branching from the initiation point formed the sclerotium initial which was a more or less spherical mass of undifferentiated, radially arranged cells. Accumulation of glycogen in cells towards the centre of the initial marked the commencement of maturation. Cells of the central (medullary) region went through a

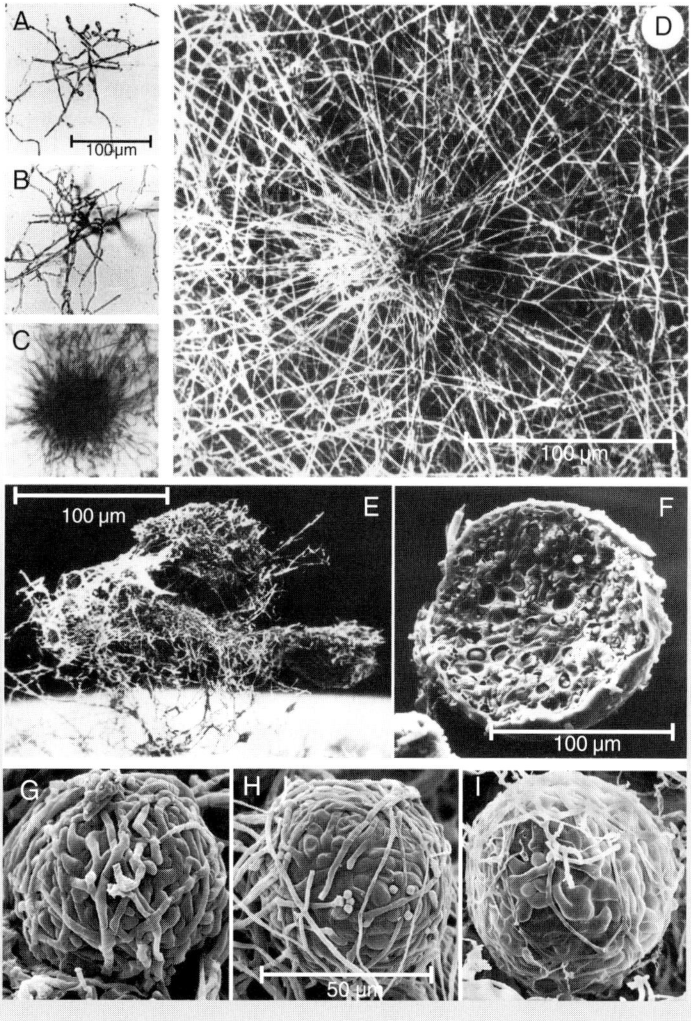

Fig. 12.6 Structure and development of fungal sclerotia. A–D, culture slide (microcosm) preparations of developing sclerotia of *Coprinopsis cinerea*. A, the earliest recognisable stage showing the centre of sclerotium initiation. B, increased size of structure as branching proceeds to form a very immature sclerotium. C, immature sclerotium showing main body with hyphae of determinate growth and the 'long hyphae' that appear to radiate outwards but are actually targeting inwards towards

differentiation process which involved first a heavy accumulation of glycogen; the glycogen was then mobilised, its reduction in concentration being exactly correlated with the formation of a thick, hyaline, secondary wall.

As the medulla developed, differentiation of localised areas of cells just within the margin of the initial occurred to form the protective rind layer. Sclerotia are a totally vegetative expression of morphogenesis. Both sclerotia and the sexual fruit bodies (ink-cap mushrooms in this case) develop from undifferentiated mycelia through an organised process of hyphal growth and branching which forms an aggregate in which cellular differentiation occurs. The early stages of this sequence are common to the processes of fruit body primordium and sclerotium formation, the cell mass becoming committed to one or other of these pathways as a result of the interaction of environmental factors like light and nutrition with the genetic factors controlling differentiation (Waters, Butler & Moore, 1975; Waters, Moore & Butler, 1975).

Caption for Fig. 12.6. continued

the developing sclerotium. **D**, scanning electron micrograph of immature sclerotium at a stage similar to that shown in **C**, which shows more clearly that the long hyphae are part of the long-distance nutrient translocation network that supports development of the sclerotium. (**A–C** are light micrographs; all images from Waters, Moore & Butler, 1975, by permission of John Wiley and Sons.) **E**, scanning electron micrograph of a group of three aerial sclerotia of *C. cinerea* in side view, relatively undisturbed and still showing the investing layer of mycelial hyphae and obvious aerial habit. (Photographed by Dr H. Waters.) The mycelial hyphae that surround the developing sclerotium, which seem so substantial in **D** and **E**, are easily lost during preparation for microscopy. **F**, scanning electron micrograph of a freeze-fractured aerial sclerotium of *C. cinerea*. This specimen is one of a collection of sclerotia that were scraped from the surface of a Petri dish culture, suspended in glutaraldehyde fixative, filtered through cotton, dehydrated into acetone, then ground in a mortar and pestle under liquid nitrogen (to fracture the sclerotia to observe their internal structure); not much of the investing layer of aerial mycelium survives this treatment. Although the chemistry is very different, the physical processes involved in this preparation are quite similar to those recorded for the preparation of *Tappania* fossil specimens (see text). (This specimen photographed by Dr F.V. Hereward.) **G–I**: scanning electron micrographs of the (smaller) sclerotia of *Byssocorticium coprophilum* (MycoBank reference Mb449580), which show more clearly that sclerotia are a 'ball' of filamentous hyphae and that constituent cells (compartments) of those hyphae differentiate as the sclerotium matures. (Images reprinted by permission from Dr J. A. Stalpers; they originally appeared on the www.mycobank.org website at this URL: www.mycobank.org/MycoTaxo.aspx?Link=T&Rec=449580#Images.)

I have seen and handled a great many '*Coprinus*' sclerotia; fresh, in actively growing cultures including microcosms, desiccated in old stored cultures with collapsed and twisted outer-layer hyphae, fixed for light microscopy and transmission electron microscopy, critical-point dried for scanning electron microscopy and, though I've never seen them after a billion years of preservation followed by dissolution into hydrofluoric acid, this first-hand experience convinces me that the *Tappania* 'vesicles' illustrated by Javaux *et al.* (2001) and Butterfield (2005) could have been at least the sclerotia of filamentous saprotrophic moulds and soil fungi (Fig. 12.6). I say 'at least' because in *Coprinopsis cinerea* the same genetic pathway produces sclerotia (as vegetative survival structures) and/or the initials/primordia of the (mushroom) fruit body depending on temperature and illumination during cultivation (Moore, 1981). So the *Tappania* 'vesicles' may also be sclerotia or the initials of ascomata or basidiomata fruit bodies. Potentially, this interpretation means that filamentous moulds able to regulate hyphal branching and hyphal interactions with sufficient finesse to assemble multicellular survival and, perhaps, reproductive structures were common and widespread anything up to 2 billion years ago.

Martin *et al.* (2003) suggested that a eukaryotic phylogenetic tree with fungi first would make sense:

> ... on the basis of available data, it seems that fungi have the broadest energy metabolic (physiological) diversity of any eukaryotic group. The fungi encompass many species with typical aerobic mitochondria, species with anaerobic mitochondria that can perform nitrite respiration, species with hydrogenosomes, species that can perform a hitherto unique feat among eukaryotes called ammonia fermentation, groups with extremely reduced mitochondria, and groups that perform methylotrophy, that is, they can live from methanol as their sole carbon and energy source, something no other eukaryotes to the authors' knowledge can. Furthermore, the fungi as a group are osmotrophs, not phagotrophs. They take up their nourishment with the help of membrane-localized importers, just like phagotrophs do, but they do not phagocytose large particles as food vacuoles. The digestion enzymes that phagocytotic eukaryotes excrete into food vacuoles, fungi excrete into their environment. The importers that phagocytotic eukaryotes use to import digest from food vacuoles reside on the plasma membrane in fungi. It is conceivable that the ***fungi as a group could have diverged from the main stem of eukaryotic evolution before proper phagocytosis had evolved.*** (Martin *et al.*, 2003, p. 197; the emphasis is mine)

Martin *et al.* (2003) based their overall tree of life on the standard three-domain model and showed the stem eukaryotes (ELCA, the Eukaryote Last Common Ancestor) as emerging from within the archaebacteria. I would adhere, as above, to the four ages of life as set out by Cavalier-Smith (2010a) but would start the age of eukaryotes about 1.5 to 2 billion years ago and amend the origin of eukaryotes as shown in Fig. 12.7.

The eukaryotic stem added the mitochondrion by enslavement of a bacterium (and perhaps added the nucleus by enslavement of an archaean, or by inheritance from LUCA, depending on the timing of divergences of prokaryotes), and evolved the endomembrane system and cytoskeletal architecture. The first stem eukaryotes probably looked a lot like primitive chytrid fungi. Single cells, of course, and still located in the biofilm where their heterotrophic lifestyle, particularly their ability to export enzymes into the biofilm matrix, enabled them to exploit nutrients sequestered by the matrix, and extract nutrients from the matrix components, as well as from dead and dying bacteria in the biofilm. To reproduce they would have converted their thallus into a sporangium that formed motile (flagellated) zoospores. The zoospores could have been released from the biofilm either by the use of matrix-digesting enzymes or, less destructively, through the thallus/sporangium restarting localised wall growth to produce a beak or neck allowing the sporangium to penetrate the biofilm to its surface. Release of motile zoospores into the fluid surrounding the biofilm allows them to distribute themselves to other biofilms or biofilm debris – there to settle and form the next generation of thalli. In this simple stem eukaryote lifestyle lays all the promise of the great diversity and evolutionary adaptation of the eukaryotic organisms we know today; this was ELCA, the eukaryotic last common ancestor. As ELCA diversified and became more widely distributed within what was still a prokaryotic world, the following features, which in the present day are characteristics of fungi (see Chapter 2), emerged in this temporal sequence:

(a) **Free cell formation**, the cytoskeletal organisation to manage vesicle and organelle trafficking and particularly the positioning of wall- and membrane-forming vesicles to enclose volumes of cytoplasm to subdivide sporangia into spores, with adoption of a chitinous cell wall, possibly as an adaptation of the ancestral actinobacterial mechanism for addition of oligosaccharides containing *N*-acetylglucosamine to surface proteins (muramopeptide wall precursors).

  (i) After this process is established, this is a potential branch point for divergence at the unicellular level to plants and heterokonts. Heterokonts are the ancestral cells of chromists (a group that today includes brown algae, water moulds, and diatoms) with one to several different types of anterior flagella; that is, all the flagella in these organisms are placed at the front of the cell. In this lineage phragmoplast formation may be left as a vestige of free cell formation, which is specifically localised at the equator of the division spindle in plants. The early cell wall became adapted to be a polymer of glucose rather

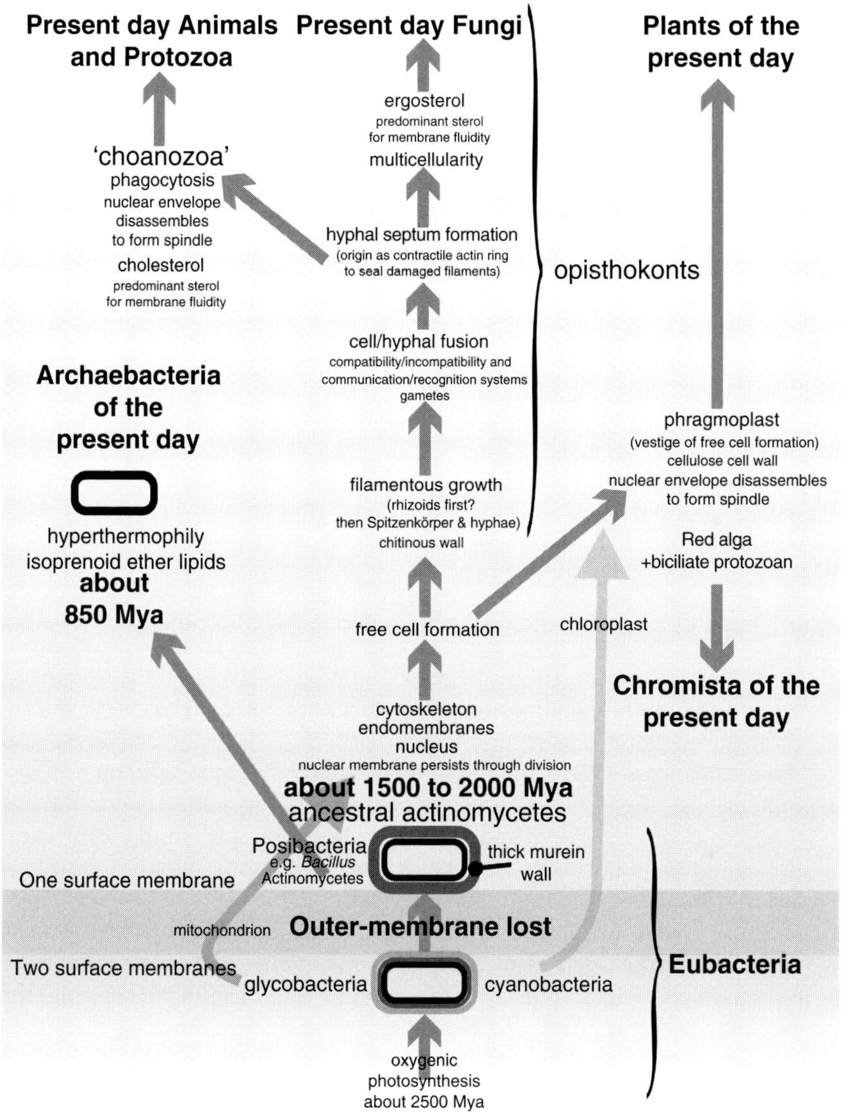

Fig. 12.7 The tree of eukaryotes. Eukaryotes diverge from actinobacterial ancestors about 2000 to 1500 mya (million years ago). The most ancient stem eukaryote (ELCA, the eukaryote last common ancestor) is considered to exhibit characteristics of primitive (chytrid) fungi. ELCA's evolution emphasises increasingly detailed management of the positioning and distribution of membrane-bound compartments (vacuoles, vesicles and microvesicles) by the filamentous components of the cytoskeleton (microfilaments, intermediate filaments and microtubules); culminating, as far as filamentous fungi are concerned, with emergence of the Spitzenkörper and apical hyphal extension. In the course of this central fungal line of evolution the other eukaryote crown groups diverged, first the plants and finally the animals.

than *N*-acetylglucosamine, possibly to economise on the demand for reduced nitrogen in an organism that is abandoning heterotrophy in favour of photosynthesis after enslavement of a cyanobacterium.

(b) **Filamentous growth**, first to make rhizoids and sporangial necks and stalks in chytrid-like opisthokont cells. Opisthokonts are the ancestral cells of both animals and fungi, their common feature being that these cells had (and still have) a single (posterior) flagellum located at the rear of the flagellated cell. Limiting wall growth to an apex even in an opisthokont cell involves creation of a coordinated production and distribution system for wall and membrane precursors and enzymes, together with a cytoskeletal delivery system and a cytoskeletal tethering system to stabilise the growing wall, weakened by insertion of new precursors, against osmotic stress (for discussion of the situation in present-day fungi see Moore *et al.*, 2011, pp 137–144; Read *et al.*, 2009, 2010; Riquelme *et al.*, 2007; Riquelme & Bartnicki-García, 2008; Steinberg, 2007; Steinberg & Schuster, 2011).

(c) **Cell fusion** primarily involves adaptation of cell wall construction functions to enable organised ***disassembly*** of two cell walls in contact, without risking osmotic stress to either cell, permitting their two cell membranes to make the two cytoplasms coextensive (see Glass *et al.*, 2004 for discussion of present-day fungi). In the ancient opisthokont cells the selective advantage of cell fusion could have been the opportunity it offers for collaboration in heterokaryons and heteroplasmons. Success could lead to the emergence of cytoplasmic (vegetative) and nuclear (sexual) compatibility/incompatibility systems (self/non-self recognition) which would allow cytoplasmically compatible cells on the one hand to exchange nuclei and form heterokaryons and on the other hand to exchange of dissimilar nuclei as a prelude to sexual reproduction and all that means for evolutionary progress.

(d) **Septum (cross wall) formation** is primarily a way of protecting the cell from the hazard of loss of cytoplasmic contents following puncture of the osmotically pressurised hydrostatic system. This is something that would also have been a hazard for the most ancient, single-celled, stem eukaryotes. There is, consequently, selective advantage in developing a contractile ring of actin as a way to seal damaged cells rapidly.

    (i) After all these processes were established at the opisthokont level, this becomes a potential branch point for divergence to animals. A divergence between chytrid-like opisthokonts that will develop into the fungal lineage on the one hand, and choanozoa-like opisthokonts on the other hand that will gradually lose the rigid wall and adapt the cytoskeletal organisation/vesicle trafficking originally used in wall synthesis and stabilisation to new functions of phagocytosis, locomotion and contractile cell division in the animal lineage.

This branch event could also have been the point in time when fungi became (possibly accidentally) fixed on ergosterol as the quantitatively predominant sterol involved with controlling membrane fluidity in contrast to the cholesterol used in animals. The chytrid-like opisthokont now becomes the 'stem-fungus' which evolves into ancestral fungi apically extending with the Spitzenkörper as the ultimate organising centre for hyphal extension and filamentous hyphal morphogenesis. Evolution of autotropism, gravitropism and other tropisms can be seen as part of this evolutionary thread, because the fundamental basis of a tropism in filamentous fungi is the directional steering of the Spitzenkörper. Cell fusion, now better called hyphal fusion, evolved to convert the otherwise radially arranged hyphae in the central regions of a maturing colony into a fully interconnected network through which materials and signals can be communicated efficiently. The selective advantage here is that the physical integration allows the vegetative mycelium to make best use of the resources its exploration has discovered. Fusion primarily involves joint adaptation of Spitzenkörper function to enable organised disassembly of two hyphal walls in contact (without risking osmotic stress to either hypha) and their two cell membranes to make the two cytoplasms coextensive. Once the process of hyphal fusion has been established as a means of enhancing the efficiency of the mycelium it could be adapted to other functions within and between mycelia. This would include the creation of multicellular structures and provision of a route for intrahyphal communication for their regulation using secondary metabolites. Emerging cytoplasmic (vegetative) and nuclear (sexual) compatibility/incompatibility systems (self-/non-self-recognition) would allow cytoplasmically compatible mycelia to exchange nuclei and form heterokaryons on the one hand and, on the other hand, to selectively exchange genetically dissimilar nuclei as a prelude to sexual reproduction and so abandon motile gametes in the fungal lineage.

Hyphal-septum (cross wall) formation, in its origins primarily a way of protecting the exploratory extending hyphal filaments from the hazard of loss of cytoplasmic contents following puncture of the hypha, became elaborated with ingressive wall formation. First, this made complete (imperforate) septa to isolate particular parts of the hyphal network (spore-forming branches, for example) and was later refined to form regularly perforate septa that allow longitudinal communication along the hypha to be maintained but, combined with a rapidly deployed septal pore plug, also save punctured hyphae from leaking to death.

The sequence of events described above is shown in summary form in Fig. 12.7. Taken together this evolutionary logic would have allowed filamentous fungi to emerge about 1.5 to 2 billion years ago as the first crown group of eukaryotes to take their place on the world stage. They developed to exploit a particular environment: the debris left by the previous 2 billion years of prokaryote growth. Although prokaryotes can also recycle debris, they need to have that debris within their local biofilm to digest it, whereas fungal hyphae can explore away from their point of origin. Above the strand lines of oceans, lakes and rivers dead and dying prokaryote microbial mats and biofilms had been tossed by storm and tempest, dried in the unfiltered rays of a brightening sun, and cracked and broken by wind and rain until covered by the detritus thrown up by the next storm – for two thousand million years. This is what awaited the first explorations of filamentous fungi; probably the first instance of an oft-repeated feature of fungal evolution, namely that fungi benefit from wide-scale extinction events.

The period 800 to 600 million years ago featured three successive virtually global glaciations (snowball Earth episodes). Cavalier-Smith (2010a, p. 127) suggests that these 'surely would have retarded early protist diversification' but I can see these episodes prompting and benefiting diversification of fungi in general and filamentous fungi in particular to exploit the death and destruction of other organisms in the same way that fungi benefited at later extinction events.

Analysis of the Permian–Triassic (P–Tr) extinction event that occurred approximately 251 million years ago (known as the Great Dying and the Earth's most severe extinction event so far) includes the quotation:

> ... sedimentary organic matter preserved in latest Permian deposits is characterised by unparalleled abundances of fungal remains, irrespective of depositional environment (marine, lacustrine [i.e. lake sediments], fluviatile [i.e. river/stream deposits]), floral provinciality, and climatic zonation. (Visscher *et al.*, 1996, quotation comes from the abstract)

Much the same is true for the Cretaceous–Tertiary (K–T) extinction of 65 million years ago, the result of a meteor collision that caused the Chicxulub crater in Mexico, which is blamed for the extinction of the dinosaurs. There was also widespread deforestation right at the end of the Cretaceous, which is assumed to be due to post-impact conditions. However, coincident with all this death and destruction of animal and plant life at the K–T boundary there is a massive proliferation of fungal fossils:

> This fungi-rich interval implies wholesale dieback of photosynthetic vegetation at the K–T boundary in this region. The fungal peak is interpreted to represent a dramatic increase in the available substrates for [saprotrophic] organisms (which are not dependent on photosynthesis) provided by global forest dieback after the Chicxulub impact. (Vajda & McLoughlin, 2004)

So it is the same story as at the other extinction boundaries: while the rest of the world was dying, the fungi were having a party!

But that Chicxulub meteor might not have had the last word on dinosaur extinction, because the massive increase in the number of fungal spores in the atmosphere of the time may have caused fungal diseases that 'could have contributed to the demise of dinosaurs and the flourishing of mammalian species' (Casadevall, 2005). A reminder, perhaps, that the fungi started the eukaryote journey by spring-cleaning the early Earth, and they've been cleaning up and modifying the planet and its biosphere ever since.

THIRTEEN

EMERGENCE OF DIVERSITY

In the story of life on Earth we have now reached the start of the great spreading out of the eukaryotes, from about 2 billion years ago. This is called the Orosirian Period (2.05 to 1.8 billion years ago). The climate of the Earth at this time is reasonably temperate (the Huronian ice age ended 2.1 billion years ago); the Sun's luminosity is at about 85% of the present-day level and the day length is about 19 hours (see Table 13.1, located at the end of this chapter). By the end of this period, bacteria are abundant and there has been sufficient prokaryotic photosynthesis for oxygen to start accumulating in the atmosphere. Geologically this is a period of intensive mountain development, which means that there were extensive expanses of terrain above sea level as well as high mountain areas and all the potential habitats that such regions represent. The period was bracketed by two major impact events: the impact 2.02 billion years ago that produced the 300 km meteor crater at Vredefort, South Africa, and the impact of 1.85 billion years ago that formed the 250 km crater at Sudbury, Ontario, Canada (you can find details in the Earth Impact Database at this URL: www.passc.net/EarthImpactDatabase/index.html).

This is the global habitat into which ELCA, the eukaryote last common ancestor, first emerged, still in its biofilm and still surrounded by prokaryotes. Those prokaryotes gave birth to eukaryotes, in the sense that ELCA diverged from a prokaryotic ancestor; but they also compete with ELCA, so to become a success ELCA must immediately compete in the life game and win. By the time ELCA emerged the prokaryotes had had at least 1.5 billion years to diversify and spread themselves throughout the world. The descendants of ELCA will compete with all of these in due time, but at the beginning ELCA was in competition only with local prokaryotes within the biofilm in which it arose. Remember that one of the rules of the life game is that revolutionary advances in evolution take place in small volumes. Two billion years ago bacteria were distributed around the globe and were so numerous that their photosynthetic activity was already modifying the atmosphere and, most dramatically at that time, precipitating the iron from the oceans as iron oxides to make the iron ore strata that humans will eventually mine for their tools. Somewhere in this world of bacteria, in an isolated but sheltered place, ELCA arose, probably in a single drop. Before we go further into a description of the evolutionary radiation from that place, I want to briefly review the most important revolutions in cell structure (or adaptations generating major evolutionary advances) that have occurred so far, so that we can remind ourselves of the rules of the life game:

(a) play the game in temperate conditions,
(b) stay in the biofilm,
(c) work in small volumes,
(d) be inventive,
(e) cooperate with those around you, but on your own terms,
(f) but compete aggressively.

So far I have referred to the following great evolutionary phases:

(a) from the prebiotic to LUCA;
(b) from the isolated first cell that is LUCA to the global diversity of bacteria, and on to the endosymbiotic collaboration leading to ELCA.
(c) ELCA gave rise to the global diversity of eukaryotes:
    (i) which drove some prokaryotes further into extreme environments (hot, cold, deep, etc.) to give rise eventually to emergence of 'archaea' extremophiles,
    (ii) while the fungi continued the theme of collaboration and stimulated the evolution of green algae, green plants and animals.

FROM THE PREBIOTIC TO LUCA

I have argued in Chapters 7, 8 and 9 that life emerged from non-enzymatic systems of autocatalytic chemical cycles in any of a range of enclosures (Chapter 10) that I have called pre-alive systems, which later acquired the capacity of enzymatically controlled metabolism where the reactions were catalysed by enzymatic RNA (ribozymes, Chapter 10); aided and abetted by clays, pyrites and crystals. The primitive Earth was a hotbed of chemical reactions with a massive range of mixed processes (regarded here as chemical experiments) taking place. Some slowly continuing and accumulating; others halting for lack of reactants but then being accelerated when mixed with others as the result of atmospheric disturbance or geological activity. I place particular stress on this prebiotic chemistry taking place in vast numbers of aerosol droplets stirred up into the atmosphere by raging storms and exploding volcanoes (Chapter 7). I suggest in Chapter 10 that life originated as a biofilm and claim the precursors and components of the first biofilms were brought together from around the globe by all those drifting aerosol droplet 'reaction vessels'. In the rainwater and seawater trickling through the roofs of volcanic caves on the spindrift-washed shore of a volcanic island in an endless shallow sea the aerosols deliver chemicals that create a slime on the volcanic sand. It is in this slime that the pre-alive systems are stabilised and the collection of coacervates, liposomes, chemotons, jeewanus, surface metabolists, chemiosmotic bubbles, mineral and crystal surfaces, and maybe others still unnamed, can react and interact in a relatively undisturbed and highly localised environment.

In this constrained environment of the slime (which, in due course, will become the first biofilm) metabolism is created by the interlinking of chemical cycles in chemotons. Different chemical cycles may come to co-operate if one produces a starter metabolite for another cycle, or if one takes a waste product from another cycle to process further, or if one cycle catalytically produces a compound that serves as a catalyst for another process. I can imagine these and many other short cycles originating as chemotons in coacervates or liposomes in individual aerosol droplets, and being brought by wind and weather to one and the same location, where in their shared smear of slime they can find their collaborative relationships. Most of the likely membrane-bound chemoton vesicles would be able to merge or combine to produce more complex 'combination chemotons' in which integrated chemical cycles ensue. This collaborative process is likely to be promoted by their localisation in the slime and is probably the first instance of the First Law of Biology, 'co-operation works', being put into practice (introduced in Chapter 2) by these pre-alive systems. Eventually, as each additional chemical cycle confers its advantages on the combined whole, this steady integration and incorporation of chemical cycles produces an overall chemoton with three subsystems:

(a) the boundary (phospholipid) membrane;
(b) the metabolic subsystem, consisting of small-molecule metabolic networks which produce all the chemical components needed for the working and reproduction of the total system, include the building blocks for membrane formation (the 'cytoplasm');
(c) the information subsystem (genetic apparatus), containing the genetic information for the whole system.

We are now getting close to something in which the chemical cycles are so complex and well integrated that the structure crosses the borderline between the inanimate and the animate; the chemoton can be considered a living prokaryotic system, LUCA.

> Although many different types of metabolism are known, which is what makes it possible for prokaryotes to live in a variety of extreme environmental conditions, each of them must have an autocatalytic metabolic network to produce the raw materials for membrane formation and to reproduce the information subsystem, i.e. the genetic subsystem. It is also true that both the boundary membrane and the information subsystem (genetic apparatus) are autocatalytic and self-reproducing, and that their co-operation results in a chemical supersystem, where the coordinated functioning comes primarily from the stoichiometric connections between the three different subsystems. Everything else, including the chemical kinetic relations, is based on these stoichiometric connections.
>
> The existence and co-operation of these three subsystems is the prior condition for the presence of life at the prokaryotic level. If any of them is absent, the system is no longer alive. In the case of viruses, only the information subsystem and, for more complex viruses, the boundary membrane are present, the metabolic subsystem is missing. Viruses cannot function by themselves they cannot grow and they cannot multiply. They can reproduce only by entering cells and forcing them to make copies to produce identical viruses. Similarly, one may make lysates from prokaryotes, i.e. material in which the cell membranes have been destroyed. These lysates can be used to synthesize biochemicals, because the metabolic subsystem still works for a time. However, these are no longer living systems; they do not have the organization of the chemotons and they cannot grow or proliferate. Their functions are not directed by an information subsystem, but only by the general rules of chemistry. Thus, with the help of the chemoton model we can define clearly the difference between living and non-living systems at the prokaryotic level. (Gánti, 2003, p. 7)

For further illustration and explanation I would refer you the end of Chapter 6, above, which features a quotation from Robert Hazen (2005, his chapter 19) that he entitles 'Three scenarios for the origin of life'.

FROM THE ISOLATED FIRST CELL THAT IS LUCA TO THE GLOBAL DIVERSITY OF BACTERIA, AND ON TO THE ENDOSYMBIOTIC COLLABORATION LEADING TO ELCA

LUCA emerged first into a single drop of slimy water (the first real biofilm). When it did first emerge, that drop, perhaps located in a crevice on the wall of a volcanic cave where it was protected from sunlight, contained the only living thing on the planet. Everywhere else was sterile at that time. Certainly, there were many other places where pre-alive systems were working through their chemical cycles in their smears of slime, and the atmosphere still teemed with the tiny aerosol reaction vessels of prebiotic chemistry; but only LUCA was alive. And at the very beginning LUCA was alone. Just one chemical supersystem chemoton that had become so complete and integrated that it satisfied the criteria and lived. It didn't have to live well; just well enough to grow and divide (replicate) occasionally. LUCA did not have to be a perfect, efficient, highly controlled living cell. It could be slow, inefficient, uncontrolled and imperfect and still be the most successful living organism in existence because it was the only living cell in existence at the time.

Because of the lack of competition, LUCA could still get by even if some of its most important processes depended on spontaneous chemical activity. For example, membranes have a natural ability to expand by insertion of monomers (= growth) and a liposome that does grow by expansion of its membrane reaches a size at which it will spontaneously fragment into smaller liposomes. In LUCA's case this is happening in a biofilm, so the fragments do not go very far and if the smaller liposomes are not working replicas of LUCA at the first attempt, the liposomes can fuse together and try again. In the previous section I stated that the LUCA chemical supersystem needs three subsystems, the membrane, the metabolic subsystem and the information (genetic) subsystem, but this does not imply or require a modern genetic apparatus; rather, all that is necessary is for LUCA's offspring to inherit a population of molecules capable of continuing LUCA's own pattern of metabolism.

The inheritance need not be exact. LUCA's offspring must have representatives of all the molecules constituting the living subsystems, but if LUCA splits into two daughter cells, the two offspring do not need a precisely apportioned 50% of those molecules. Cell division for LUCA will be effective providing the components of its subsystems are distributed at random to its offspring. The rest – that is, the efficiency and the control (the perfection) of cell division – would come with time by a combination of genetic drift and Darwinian selection. Indeed, inaccuracy in the distribution of components during divisions to form daughter cells could have been, at this early stage, the equivalent of present-day mutational genetic drift, generating variation on which Darwinian selection could operate.

In modern evolutionary theory genetic drift is the change over time in gene frequencies of a population due entirely to random events rather than natural selection. Natural (or Darwinian) selection acts on the relative reproductive success of the competing variants; it ensures survival of the individuals and their progeny that prove themselves best equipped to adapt to the environmental conditions encountered by the population. In large populations, which have representative distributions of genes and in which truly random sampling (in modern populations that usually means 'random mating') is effective, the level of genetic drift is usually negligible and it plays a minor role in evolution, at best. In contrast, small populations may not have a representative distribution of all possible genes with the result that the sample is non-random and genetic drift results in some genes becoming more common while others become less common from generation to generation.

Modern evolutionary theory is based on the reproductive processes of present-day organisms (DNA, chromosomes, mitosis, meiosis, sexual reproduction), but neither genetic drift nor Darwinian selection is logically dependent on such reproductive processes; indeed, Darwin knew nothing about DNA, chromosomes, mitosis or meiosis when he developed the idea of natural selection. It is entirely appropriate to apply these theoretical interpretations to LUCA and its progeny providing we remember that this does not imply that we are assigning to LUCA a genetic architecture as mature as those of present-day bacteria. Since LUCA started with a population size of one, it is reasonable to expect that genetic drift acting alone would be an important, perhaps the only, way of generating diversity in those primeval days.

> ... life began and flourished without the benefit of exact replication, then it is appropriate to assume that genetic drift remained strong and natural selection remained relatively weak during the early exploratory phases of evolution. But this is not to say that Darwinian selection had to wait until life learned to replicate exactly ... Darwinian selection is unavoidable as soon as inheritance begins, no matter how sloppy the mechanism of inheritance may be. (Dyson, 1999, p. 20)

As soon as LUCA split (replicated) for the first time it had the potential of producing diversity in its offspring, and a diverse population will be subject to genetic drift and natural selection.

The earliest living things (LUCA and its progeny) were still within their original biofilm. They were heterotrophs, using readily available nutrients which were absorbed by the biofilm from its surroundings (atmosphere, rainwater, seawater) or were synthesised by the pre-alive chemotons that still coexisted with LUCA. The nature and distribution of these nutrients must have varied throughout the biofilm, and this variation could have selected for different metabolic specialists among LUCA's diverse progeny. As environmental nutrients were exhausted, chemical catalysts (ribozymes; see Chapter 10) were developed for release into the environment to degrade the polymers accumulated by hundreds of millions of years of abiotic chemical reactions, some of which would be components of the primitive biofilm matrix.

Some of the ribozymes were not fully released, but extruded through the membrane, anchored internally by hydrogen bonding to a parking RNA, to be rewound by hydrogen bonding when it had picked up nutrient. This process provides a metabolic selection pressure for the evolution of greater specificity for nutrient (to improve the scavenging function) and greater specificity between the two RNAs (that is, the emergence of something like specific codon–anticodon purine-to-pyrimidine hydrogen bonding) to allow improved placement of the ribozyme (Chapter 10). Note also the possibility also discussed in Chapter 10 that pseudo-peptides conferring some local feature on the membrane, such as adhesion, hydrophobicity, ion selection and/or nutrient binding, could have been made by lining up amino acid-binding ribozymes on a parking RNA; and that this suggests a metabolic selective advantage for something that resembles the specific 'messenger-RNA/transfer-RNA/amino acid' relationships of the present-day transcription/translation and protein-synthesis machinery.

The selection pressure associated with exhaustion of environmental nutrients could have also favoured progressive evolution for any of LUCA's offspring that experimented with photochemical reactions using jeewanus-type chemotons coexisting in the biofilm. In due course these may have become sufficiently proficient at photosynthetic fixation of carbon and nitrogen to support the metabolism of those descendants of LUCA that drifted out of the shadows and into the sunlit regions of their biofilm.

By these means both the diversity and distribution of the first bacteria increased. In the 1.5 billion years following the first highly localised emergence of LUCA the diversity and distribution increased to a global covering of prokaryotes until they had successfully penetrated every temperate habitat on Earth. They would then have invaded successively more hostile habitats; in view of likely local conditions, I would anticipate explorations of widened temperature tolerance: high temperatures in volcanic waters and low temperatures in elevated mountain regions. Saline muds must have been another common habitat fringing the ocean, lakes or river estuaries in which extreme salt tolerance was likely of adaptive value. The diverse world of the prokaryotes was broadly established during that 1.5 billion years (Figs. 11.2, 11.3) – the range of different morphological types: spheroidal, rod-like, thread-like, or spiral; motionless or motile. Most will have had solid cell walls, and the majority were still in biofilm communities, the pelagic/planktonic existence being itself a specialisation of the minority. In a sense, of course, the biofilms also evolved because their nature depends on the activities of their living components. Those cells would adapt their contributions to the biofilm matrix as they became specialised: protective pigments for sunlight-exposed locations, more adhesives to stabilise a biofilm in flowing water, charged polysaccharides and proteins to sequester mineral ions from the fluid, etc. (see Table 10.1). Add to all this biodiversity a further diversification of metabolic types, ranging from sulphur bacteria to methane-producing bacteria.

All prokaryotes were initially anaerobic, but, thanks to the first photosynthetic bacteria, the first free oxygen appeared in the oceans and atmosphere about 2.5 billion years ago. For a long time most of this oxygen was titrated out as iron salts dissolved in the oceans were precipitated to the ocean floor as iron oxide muds that eventually created the iron ore formations from which humans would extract the iron for their technologies (Dobson & Brodholt, 2005). The photosynthetic organisms producing the oxygen are members of a biofilm community, but anaerobic organisms are poisoned by oxygen; consequently the rest of the community (and the photosynthesisers themselves since they are most immediately exposed to the oxygen) will be subjected to a selection pressure in favour of some way in which the oxygen can be detoxified. One way in which this can be done is to transfer two electrons and two protons to each oxygen atom to make water. Now, if an enterprising primitive bacterium could find one of those chemiosmotic bubbles of iron sulphide, the sort that precipitate in the cool alkaline water around shallow hydrothermal volcanic vents (see Chapter 6), it might be able to put the proton gradient they generate to good use in detoxifying oxygen; and this just might be the start of something big. It seems to have taken a couple of hundred million years, but I guess that must be close to what happened because about 2.2 billion years ago organisms with mitochondria capable of aerobic respiration appear (organisms corresponding to ELCA (the eukaryotic last common ancestor) or a close descendant of ELCA), which implies that bacteria capable of aerobic respiration appeared a little time before.

Just over 2 billion years ago, then, bacteria were abundant, widespread and very diverse. Successful biofilm communities existed in all habitats around the globe containing vigorous mixed populations of heterotrophs and autotrophs, anaerobes and aerobes. Their biofilm existence established the advantages of collaboration, but their physically close containment in the biofilm provided them with one further and more radical opportunity: for some endosymbiotic collaborations leading to combination organisms and eventually ELCA. Lynn Margulis has been the chief advocate for the theory that the main organelles of eukaryotes – mitochondria and chloroplasts – originated as endosymbiotic aerobic bacteria or cyanobacteria, respectively (Margulis, 2004).

> The theory implies that the original eukaryote was an anaerobic fermenting organism and that an aerobic bacterium established itself as a symbiont within this protoeukaryote. The symbiont could then utilize metabolites of the host for its own aerobic metabolism. It is possible that the original symbiont was photosynthetic, like modern purple non-sulphur bacteria and that it later lost its capability of photosynthesis. (Fenchel, 2002, chapter 8, p. 85)

This model for the origin of mitochondria and chloroplasts is now well documented and generally accepted, and it is thought that chloroplasts have evolved independently on several occasions in various present-day groups of eukaryotes. Fenchel explains:

> Both types of organelles reproduce by division (are replicators), have their own bacteria-type chromosome, an apparatus for transcription and translation, and bacteria-type ribosomes. Evolutionary trees based on the sequence of rRNA genes ... unambiguously place mitochondria and chloroplasts among the α-group of the proteobacteria, and the cyanobacteria, respectively. (Fenchel, 2002, chapter 8, p. 84)

Fenchel (2002, pp. 84–86) also points out that there are many recent examples of endosymbiotic prokaryotes inside present-day eukaryotic cells functioning in ways similar to mitochondria and chloroplasts. It is usually supposed that the endosymbiosis arose when one bacterial cell ingested, but failed to digest, a proteobacterial cell that was to become the mitochondrion, and subsequently the same fate befell a cyanobacterial cell, ingested but not digested, that also became a symbiont. Margulis (2004) derives other eukaryotic features (the eukaryotic flagellum and the centriole or spindle pole bodies; see Chapter 2) from similar 'failure to digest'/endosymbiotic relationships. Eukaryotic flagella (which are probably best called cilia) differ quite fundamentally from bacterial flagella both in structure and in the way they operate; so the two structures are not related by descent. Margulis also derives the eukaryotic nucleus by combining the genomes of at least the two different prokaryotes involved in creating the mitochondrial combination. Although there is little doubt about the origins of mitochondria and chloroplasts as prokaryote endosymbionts there is less compelling evidence for these others, and I discuss the major theories that have been proposed to explain the origin of the eukaryotic nucleus in Chapter 11. In discussing how different pre-alive systems co-operated to generate the first prokaryotes ('From the prebiotic to LUCA', above in this chapter), I suggested that process to be the initial implementation of my First Law of Biology ('co-operation works', introduced in Chapter 2), and I suppose that life in a biofilm must be classed as the second implementation. I believe that the origin of mitochondria and chloroplasts as prokaryote endosymbionts is the third revolutionary implementation of this principle.

Fenchel (2002, pp. 86–91) discusses how symbionts evolve to organelles, explaining first that there are shades of meaning to the term symbiosis. It is often used to describe a co-operation which is mutually beneficial to the parties involved. More often, though, biologists use the term mutualism to describe such relationships; in formal terms, mutualisms are co-operations that increase the fitness of both members. This leaves the term symbiosis meaning simply that two types of organisms live in a physically close co-operation without implication about who profits from the association. There is clear advantage for the host cell because presence of the symbiont opens new directions for evolutionary adaptation. In a sense the host cell is parasitising the symbiont, using it to out-compete relatives of the host cell that do not have symbionts. By contrast, if the relationship is stabilised and the symbiont becomes an organelle, then the host is providing the symbiont with an environment tailored to its needs and removing the symbiont from the competitive environment of its free-living relatives. The intracellular environment is still competitive, however, in the sense that selection will operate in favour of changes (mutations) in the symbiont that benefit the host cell, whereas symbiont mutations that are unfavourable to the host will be selected against. There are several examples among present-day organisms of endosymbiotic associations in which present-day prokaryotes or eukaryotes live inside the cells of other present-day eukaryotes. Fenchel (2002, pp. 86–91) describes some such examples, suggesting that they indicate that endosymbiosis has been a recurrent theme in evolution over the past 2 billion years:

> Altogether the examples presented in this section show that organelle evolution from endosymbiosis is not something that took place only 1.5–2 billion years ago, but that evolution through chimera formation between eukaryotes and prokaryotes still happens. The origin of chloroplasts and of mitochondria had an enormous effect on the biosphere. In the extant biosphere most of the light-driven fixation of CO_2 is due to chloroplasts in eukaryotes (plants on land and algae in the sea) and a very large share of respiratory O_2 reduction is due to the descendants of purple bacteria that became mitochondria. (Fenchel, 2002, chapter 8, pp. 90–91)

ELCA GAVE RISE TO THE GLOBAL DIVERSITY OF EUKARYOTES

Just as LUCA formed in a single drop, so too did ELCA. The eukaryote last common ancestor, the stem eukaryote, first emerged within and then radiated from a single very small volume of biofilm on some prokaryote-covered surface. Unlike LUCA, it emerged into a world already dominated by other organisms (its bacterial neighbours) with which it must immediately compete. In order for ELCA to survive, let alone give rise to the global diversity of eukaryotes the world now enjoys, the structure of ELCA must have given it an immediate advantage over those bacterial neighbours. It probably had an advantage in cell size. The most commonly found bacterial cells today typically generally range between 0.2 and 2 μm diameter, whereas eukaryotic cells are mostly larger, in the overall size range of about 10 to 100 μm diameter. There are exceptions to both these ranges, but generally speaking, modern eukaryotic cells are typically ten times the size of modern prokaryotic cells.

Modern, fully adapted, eukaryotic cells contain a range of space-requiring organelles (mitochondria, nucleus, cytoskeleton, centrioles/spindle pole bodies, endoplasmic reticulum, Golgi vesicles and vacuoles of various sorts, and often flagella) that explain why they must be larger than modern bacteria. Although few of these were shared by ELCA, on the principle that what is general today is likely to be ancestral, it is reasonable to assume that the stem eukaryote 2 billion years ago was about ten times the size of its bacterial neighbours. This large unicellular, aerobic, heterotroph (it has not yet ingested a cyanobacterium), which is possibly motile (though it may only have motile zoospores), with a strong polysaccharide-based cell wall is the stem eukaryote. It is sufficiently competitive to win the life game against its prokaryotic neighbours and could have been a primitive chytrid fungus exploiting, by recycling, the debris left by earlier prokaryote growth (Chapter 12).

While its descendants evolved and radiated during the next 2 billion years to a new global domination, success of the eukaryotes drove the prokaryotes further into the extreme of habitats on Earth – hot, cold, deep, etc. – and eventually gave rise to emergence of that group of extremophiles: hyperthermophilic, hyperpsychrophilic and hyperbarophilic bacteria which are unfortunately called 'archaea' though they evolved so recently (Fig. 12.7) (Cavalier-Smith, 2006, 2010a). As an aside, I would also add that the corrosive nature of an oxygen atmosphere means that oxygen-using (aerobic) organisms on Earth should also be considered extremophiles. The evolution of oxygen-using organisms is an extreme development that can only occur as a consequence of photosynthetic/water-photolysing living things producing an oxygen-rich atmosphere in the first place. Consequently, an essential first step in the evolution of efficient living organisms is that one of the first groups of living things to arise was so successful in competing with its neighbours that it poisoned the atmosphere of its home planet after changing the chemical nature and distribution of the fourth most common element in the Earth's crust (iron; which is the most common element, by mass, in the planet as a whole, is only number four in the crust because there is so much in the Earth's core). A dramatic example of the truism that living things modify their own environment by the act of living.

The eukaryotic stem possessed the mitochondrion and nucleus, which together characterise the eukaryotic grade of cell structure, and rapidly evolved those other characteristics of eukaryotes: the endomembrane system and cytoskeletal architecture. In Chapters 2 and 12 I suggest that endomembranes and cytoskeleton evolved in tandem with the overall selective advantage of continually improving management of vesicle transport and placement being expressed in evolution of features that are characteristics of fungi in the present day:

(a) free cell formation;
(b) filamentous hyphal growth;
(c) hyphal/cell fusion;
(d) hyphal-septum (cross wall) formation.

From this evolving stem, first plants and then animals diverged as unicellular progenitors of their respective kingdoms, leaving the chytrid-like stem eukaryote to create the fungal kingdom (Fig. 12.7).

Completion of the sequence of events shown in Fig. 12.7 allowed filamentous fungi to emerge about 2 to 1.5 billion years ago. By their very nature, filamentous fungi are multicellular organisms. Even the most primitive of present-day fungi, which only use cross walls (septa) in their hyphae to separate spore-forming or damaged hyphae from the rest of the mycelium, nevertheless differentiate their filamentous hyphae into physiologically distinct regions. The majority of filamentous fungi, in which category I include all of the ascomycetes (moulds, mildews, morels and cup fungi) and all of the basidiomycetes (mushrooms, toadstools, boletes, polypores, rusts and smuts), divide up their hyphae into linear chains of hyphal cells.

In the present day, eukaryotes are distinct from most prokaryotes on the planet by being able to create multicellular aggregations which have a distinctive pattern and a lifestyle of their own.

> The multicellular organisms represent the next major transition in the evolution of life (after the origin of the first prokaryotic cells and the origin of the eukaryotes). Three groups: fungi, plants, and animals have evolved independently, but they remain relatively close to each other on the eukaryote phylogenetic tree ... The fungi descend from a group of fungi-like protists (the chytrids), the plants from green algae, and the animals from a group of flagellates (choanoflagellates). Green algae, red algae, and brown algae are often (with good reason) included among the multicellular organisms; they also form large cell colonies that include different functional types of cells and many of these algae grow to a quite large size. Finally, many representatives of various protist groups (and even some bacteria) produce cell colonies that share some features with true multicellular organisms. (Fenchel, 2002, chapter 9, p. 92)

When most people think or write about a multicellular organism, they tend to think or write about a multicellular animal, a metazoan. A few people, gardeners or children of gardeners, may think or write about plants – roses, or rockcress perhaps, or redwoods. Some people believe that multicellular organisms appeared relatively late in the history of Earth, but that's because of their animal/metazoan fixation. Fungi, plants and animals diverged from each other in evolution at a unicellular stage (as described above), so each kingdom entered the multicellular revolution independently of the others, and at times to suit themselves.

Multicellular fungal hyphae were the first really successful multicellular eukaryotes, 1.5 billion years ago, so it was the fungi that began the eukaryote's tendency to multicellularity, first working out how to make and use multicellular balls of hyphae (Fig. 12.6), then evolving the mechanisms to control those key features of multicellularity (known as developmental biology or morphogenesis): cell differentiation, patterns of tissue distribution and establishment of body plans that feature in the development of multicellular structures like fungal fruit bodies (Fig. 12.3D) (Moore, 2005).

The underlying logic and principles of developmental biology appear to be the same in fungi, plants and animals. In all three kingdoms cells in different places in any sort of multicellular structure have to carry out different tasks and to do this must be caused to follow different routes of specialisation, which means that they follow different genetic programmes. The cells have to be instructed to differentiate in this way, and the instructions need to include where and when the differentiation should take place as well as what sort of differentiation. Consequently, there is a need for highly specific genetic control within the cells and equally specific intercellular signalling to make sure the signal is delivered and acted upon by the right cell, in the right place at the right time. Many of the signalling molecules that control animal developmental biology have been identified, as have some of the most important regulators of plant development. In these days of readily accessible genome databases it is possible to plan comprehensive searches of the sort 'does this animal developmental sequence exist in any plant genome?' and 'does this animal developmental sequence exist in any fungal genome?' and then go back and ask the same question about all known plant developmental sequences.

When this sort of survey was done (Moore & Meškauskas, 2006) the only highly similar matches found were between sequences involved in basic cell metabolism or essential eukaryotic cell processes. There were no between-kingdom similarities in the management processes that regulate multicellular development. None of the animal sequences occurred in plants or fungi; none of the plant sequences occurred in animals or fungi. In the present day the crown group of eukaryotic kingdoms control and regulate their developmental processes in very different ways. Unfortunately, we know nothing about molecular control of multicellular fungal developmental biology; it's one of those areas, like most of the characteristic cell biology of filamentous fungi, which is still opaque because of lack of research effort.

ELCA was probably very similar to what would today be called a chytrid fungus (a member of the phylum Chytridiomycota), and chytrids are still crucially important members of Kingdom Fungi in the present day. Even after 2 billion years service! Today, the Chytridiomycota are the only true fungi that are aquatic and have actively motile spores. All other fungi are terrestrial, or, if occurring in aquatic environments, have been derived from terrestrial fungi. In fact, fungi were probably the first organisms to venture onto the land, and they are now found in all terrestrial habitats, often in association with other organisms. This last statement is a particular feature of fungal evolution; from the very beginning they have worked in co-operation with other organisms. At the very beginning they formed lichens with cyanobacteria (Fig. 12.3G and H). Today, about 20% of all known fungi are involved in lichens, which are usually associations between a fungus and a green alga, although the fungus can survive independently in nature. Some lichens are tripartite associations involving both a cyanobacterium and a green alga. Lichens are combination organisms that bring together the best features of at least two and potentially three kingdoms: the ability of fungi to digest externally and release nutrients from even the most refractory materials, the ability of algae to fix carbon dioxide, and/or the ability of cyanobacteria to fix carbon dioxide and nitrogen. In the present day lichens are extremely resistant to environmental extremes and are pioneer colonisers on rock faces, tree bark, walls, roofing tiles, etc., as well as early colonisers of exposed terrestrial habitats, like mountain scree and arctic tundra. They would have been able to colonise primeval terrestrial habitats soon after such habitats were formed more than 1.5 billion years ago. To put that date into context, the oldest known fossils of animals are about 600 million years old and vascular plant fossils appear in rocks 500 million years old. Flowering plants go back in time only about 120 million years, this being about the time that Africa and India separated from Antarctica (Table 13.1).

Another fungal co-operation that shaped the biosphere of Earth is the mycorrhiza; a mutualistic interaction between fungi and plant roots that developed very early in the process of colonisation of the terrestrial environment. Mycorrhizas originated over 450 million years ago (Fig. 12.4). Now more than 6000 fungi are capable of forming mycorrhizas and at least 95% of vascular plants of today have mycorrhizas associated with their roots. Enhanced growth of the plant host occurs mainly because the fungus improves phosphate availability to the plant by digesting insoluble polyphosphates in the soil, but plant-to-plant transfer of nutrients (including water) can occur via the fungus, and in extremely harsh conditions (e.g. winter in upland regions) the mycorrhiza may support the host with carbon and nitrogen nutrients acquired by digesting proteins in the soil. Normally, though, the mycorrhizal fungus takes photosynthetically produced carbohydrates from the plant host. All climax vegetation, which is the long duration equilibrium state of ecosystems on Earth, such as tropical evergreen forests, temperate forests, tundras, savannahs and grasslands, are dependent on mycorrhizas.

Fungi also co-operate with animals. Ants evolved the ability to cultivate fungi about 45–65 million years ago. The ants actively inoculate their nest with a specific fungus and then cultivate it by providing it with pieces of leaves, pruning the hyphae and removing intruder fungi. As a reward, the fungus provides bundles of specialised hyphae that the ants use as a food source. The ants are engaged in an agricultural activity; they collect fresh leaf biomass to convert it to compost in order to cultivate a particular fungus that then provides the main food source for the nest. The demand for leaf material as the colony grows is enormous and in the tropical rainforests of Central and South America, leaf-cutter ants are the dominant herbivores. That 'dominant' label includes the humans of the forest. Around 50 agricultural and horticultural crops and about half that number of pasture plants are attacked. It has been calculated that leaf-cutting ants harvest 17% of total leaf production of the tropical rainforest. Nests located in pastures can reduce the number of head of cattle the pasture can carry by 10–30%; obviously, leaf-cutter ants are considered a pest as they are second only to human farmers in their ability to destroy tropical forest. The combination of a top-of-the-range social insect with a top-of-the-range fungal plant-litter degrader seems to be the key to this success. The social insect has the organisational ability to collect food material from a wide radius around its nest; but the extremely versatile biodegradation capability of the fungus enables the insect to collect just about anything that's available. Schultz & Brady (2008) point out that agriculture is a specialised symbiosis that is known to have evolved in only four animal groups: ants, termites, bark beetles and humans.

And human agriculture is also dependent on fungi; here's the story, judge for yourself. The story is about the symbiotic relationship between ruminants and the chytrid fungi in their guts that enable the animals to incorporate difficult-to-digest grasses into their diet. The efficiency of ruminant digestion, together with expansion of the grasslands, gave the ruminants the opportunity to become the dominant terrestrial herbivores throughout the world in the most recent epochs. However, the story starts long before the evolution of grasses because the chytrid fungi were the first to diverge about 1.5 billion years ago. Consequently, these fungi have existed on Earth, presumably as saprotrophs in anaerobic niches such as muds and stagnant pools, since before either grasses or herbivorous animals of any sort evolved about 60 to 20 million years ago.

Chytrids are 'probably the most common microbial element' in the Devonian Rhynie Chert, which is 410 million years old (Taylor *et al.*, 2006; and see Fig. 12.3C). From that time onwards the fungi were abundant, so any browsing animal that feasted on the community of plants similar to that represented in the Rhynie Chert would have got a mouthful of saprotrophic microfungi along with their salad. The first true herbivores were most probably fruit and seed eaters (frugivores) because the starch, protein and fats stored in fruits and seeds can be more easily digested than the plant fibres in foliage. It is argued that the evolution of large size was a prerequisite for the exploitation of leaves because of the need for a long residence time in the gut for fermentation to extract sufficient nutrients from foliage and herbage (Mackie, 2002). In the Cretaceous (about 100 million years ago; Table 13.1) dinosaurs probably occupied the herbivorous niche although they were still browsers; true grazing animals appeared much later in the Miocene (around 20 million years ago) with the radiation of grassland-forming grasses of the plant family Poaceae. Plant-eating mammals during the late Cretaceous and early Palaeocene (say, 80 million years ago) were physically small frugivores; mammals did not become herbivores until the middle Palaeocene (60 million years ago). Herbivore browsers first appeared in the middle Palaeocene but they did not become major components of the fauna until the late Eocene (40 million years ago). The earliest herbivores were probably large, ground-dwelling mammals, reaching their dietary specialisation by evolution from large, ground-dwelling frugivores or by a major size increase from small insectivorous ancestors (Mackie, 2002).

Grass-dominated ecosystems, including steppes, temperate grasslands, and tropical–subtropical savannahs, play a central role in the modern world; these ecosystems evolved during the Cenozoic, about 60 million years ago (Jacobs, 2004; Strömberg & Feranec, 2004). The grasses that dominate the semi-arid savannah (called C_4 grasses) appeared a little later; they can photosynthesise more efficiently in the higher temperatures and sunlight encountered by savannah grasses because they use water more effectively and have biochemical and anatomical adaptations to reduce photorespiration. There is a good argument for the evolution of the ruminants being driven by the development and expansion of savannah and steppe grasslands in Africa and Eurasia (Bobe & Behrensmeyer, 2004). The appearance of grasslands in Africa and Eurasia during the Eocene epoch, and their subsequent spread during the Miocene, saw the Artiodactyls (even-toed ungulates) begin to dominate over the Perissodactyls (odd-toed ungulates). A credible hypothesis for the evolution of rumination in artiodactyls is that it represented a joint adaptation to increasing aridity of the local environment due to climatic cooling and drying.

The Eocene climate was humid and tropical, and is likely to have favoured browsers and frugivores (and hindgut fermentation). With the onset of the Oligocene, the climate became generally cooler and drier, and this trend persisted throughout the Tertiary. This increasing aridity, coupled with high sunlight exposure in the equatorial zone, would have favoured the C_4 grasses, and as they became the dominant vegetation the emphasis in herbivore evolution would be to increase the efficiency of the fermentation of the more fibrous plant material. Selection pressure favoured rumination (Bobe & Behrensmeyer, 2004; Mackie, 2002); and rumination depends on the activity of anaerobic chytrid fungi in the animals' digestive system. The expansion of grassland at the expense of Miocene forests created conditions that were favourable for the evolution of artiodactyls that could survive aridity and exploit grassland vegetation; changes in the environment drive major evolutionary events, and in this case major changes in bovid abundance and diversity were caused by dramatic climatic changes affecting the entire ecosystem. The artiodactyls became the most abundant and successful order of current and fossil herbivores, with about 190 species living today.

An added twist to the story is that the emergence of the genus *Homo* in the Pliocene of East Africa (Table 13.1) also appears to be broadly correlated in time with the advent of these same climatic changes and the introduction of the ecosystems they brought about (Bobe & Behrensmeyer, 2004). Grasslands currently represent 25% of the vegetation cover of planet Earth; and this family of plants (Poaceae) is the most important of all plant families to human economy as it includes all of our staple cultivated food cereal grains.

(a) The grasses owe their success to their mycorrhizal fungi that enabled them to cope with the environmental pressures plants faced during the evolution of the savannah grasslands of East Africa.
(b) The artiodactyls owe their success wholly to their symbiotic relationship with their anaerobic rumen chytrid fungi.
(c) Humans found their staple cereal foods among the Poaceae and their main food animals among the ruminants.

That's why I suggest it's fair to say that fungi co-operate with humans, and it involves the oldest of fungi, the chytrids. They waited nearly 1.5 billion years to give human evolution a shove in the right direction. Thank fungus for that!

Table 13.1. Geological and biological timeline of the Earth from 2500 million years ago (mya) to the present day

- Proterozoic Eon (2500 to 542 mya)
  - Palaeoproterozoic Era (2500 to 1600 mya)
    - Siderian Period (2500 to 2300 mya)
      - Stable continents first appeared.
      - 2500 mya: First free oxygen is found in the oceans and atmosphere.
      - 2400 mya: Great Oxidation Event, also called the Oxygen Catastrophe. Oxidation precipitates dissolved iron creating banded iron formations (Dobson & Brodholt, 2005).
      - Anaerobic organisms are poisoned by oxygen.
      - 2400 mya: Start of Huronian ice age.
    - Rhyacian Period (2300 to 2050 mya)
      - 2200 mya: Organisms with mitochondria capable of aerobic respiration appear.
      - 2100 mya: End of Huronian ice age.
    - Orosirian Period (2050 to 1800 mya)
      - Intensive mountain development (orogeny).
      - 2023 mya: Meteor impact, 300 km crater Vredefort, South Africa (reference: Earth Impact Database).

Table 13.1. (*cont.*)

2000 mya: Solar luminosity is 85% of current level.
Oxygen starts accumulating in the atmosphere.
1850 mya: Meteor impact, 250 km crater Sudbury, Ontario, Canada (reference: Earth Impact Database).
Statherian Period (1800 to 1600 mya)
Complex single-celled life appeared. Abundant bacteria.
Mesoproterozoic Era (1600 to 1000 mya)
Calymmian Period (1600 to 1400 mya)
Photosynthetic organisms proliferate.
Oxygen builds up in the atmosphere above 10%.
Formation of ozone layer starts blocking ultraviolet radiation from the Sun.
1500 mya: Eukaryotic (nucleated) cells appear.
Ectasian Period (1400 to 1200 mya)
Green algae (Chlorobionta) and red algae (Rhodophyta) abound.
Stenian Period (1200 to 1000 mya)
1200 mya: Spore/gamete formation indicates origin of sexual reproduction (Butterfield, 2000).
1100 mya: Formation of the supercontinent Rodinia.
Neoproterozoic Era (1000 to 542 mya)
Tonian Period (1000 to 850 mya)
1000 mya: Multicellular organisms appear.
950 mya: Start of Stuartian–Varangian ice age.
900 mya: Earth day is 18 hours long.
The Moon is 350 000 km from Earth (Bills & Ray, 1999).
Cryogenian Period (850 to 630 mya)
750 mya: Break-up of Rodinia and formation of the supercontinent Pannotia.
750 mya: End of last magnetic reversal.
650 mya: Mass extinction of 70% of dominant sea plants due to global glaciation ('Snowball Earth' hypothesis).
The Moon is 357 000 km from Earth (Bills & Ray, 1999).
Ediacaran (Vendian) Period (630 to 542 mya)
590 mya: Meteor impact, 90 km crater Acraman, South Australia.
580 mya: Soft-bodied organisms developed: jellyfish, *Tribrachidium* and *Dickinsonia* appeared.
570 mya: End of Stuartian–Varangian ice age.
550 mya: Pannotia fragmented into Laurasia and Gondwana.
Phanerozoic Eon (542 mya to present)
Palaeozoic Era (542 to 251 mya)
Cambrian Period (542 to 488.3 mya)
Abundance of multicellular life.
Most of the major groups of animals first appear.
Tommotian Stage (534 to 530 mya)
Animals with shells appeared.
Solar brightness was 6% less than today.

Table 13.1. (*cont.*)

- Ordovician Period (488.3 to 443.7 mya)
  - Diverse marine invertebrates, such as trilobites, became common.
  - First vertebrates appear in the ocean.
  - First green plants and fungi on land.
  - Fall in atmospheric carbon dioxide.
  - 450 mya: Start of Andean–Saharan ice age.
  - 443 mya: Glaciation of Gondwana.
  - Mass extinction of many marine invertebrates. Second largest mass extinction event; 49% of genera of fauna disappeared.
- Silurian Period (443.7 to 416 mya)
  - 420 mya: End of Andean–Saharan ice age.
  - Stabilisation of the Earth's climate.
  - Coral reefs appeared.
  - First fish with jaws – sharks.
  - Insects (spiders, centipedes), and plants appear on land.
- Devonian Period (416 to 359.2 mya)
  - Ferns and seed-bearing plants (gymnosperms) appeared.
  - Formation of the first forests.
  - Earth day is ~21.8 hours long.
  - Wingless insects appeared on land.
  - 375 mya: Vertebrates with legs, such as *Tiktaalik*, appeared.
  - Atmospheric oxygen level is about 16%.
  - First amphibians appear.
  - 374 mya: *Mass extinction of 70% of marine species. This was a prolonged series of extinctions occurring over 20 million years.
  - Evidence of anoxia in oceanic bottom waters, and global cooling. Surface temperatures dropped from about 93 °F (34 °C) to about 78 °F (26 °C).
  - 359 mya: Meteor impact, 40 km crater Woodleigh, Australia.

Carboniferous Period (359.2 to 299 mya)

- Mississippian Epoch (359.2 to 318.1 mya) (Lower Carboniferous)
  - 350 mya: Beginning of Karoo ice age.
  - Large primitive trees develop.
  - Forests consist of ferns, club mosses, horsetails and gymnosperms.
  - Oxygen levels increase.
  - Vertebrates appear on land.
  - First winged insects.
  - Seas covered parts of the continents.
  - Animals laying amniote eggs appear (318 mya).
- Pennsylvanian Epoch (318.1 to 299 mya) (Upper Carboniferous)
  - 310 mya: First reptiles.
  - Atmospheric oxygen levels reach over 30%.
  - Earth day is ~22.4 hours long.
  - The Moon is 375 000 km from Earth (Bills & Ray, 1999).
  - Giant arthropods populate the land.

Table 13.1. (*cont.*)

Transgression and regression of the seas caused by glaciation.
Deposits of coal form in Europe, Asia and North America.
Permian Period (299 to 251 mya)
275 mya: Formation of the supercontinent Pangea.
Conifers and cycads first appear.
Earth is cold and dry.
Sail-backed synapsids such as *Edaphosaurus* and *Dimetrodon* appeared.
260 mya: End of Karoo ice age.
251 mya: *Mass extinction (Permian–Triassic).
Earth's worst mass extinction eliminated 90% of ocean dwellers and 70% of land plants and animals.
Possible 480 km-wide meteor crater in the Wilkes Land region of Antarctica (von Frese *et al.*, 2009).
Period of great volcanism in Siberia releases large volume of gases (carbon dioxide, methane and hydrogen sulphide) (Ward, 2006).
Oxygen levels dropped from 30% to 12%.
Carbon dioxide level was about 2000 ppm.
Mesozoic Era (251 to 65.5 mya)
Triassic Period (251 to 199.6 mya)
Break-up of Pangaea starts.
Survivors of P–T extinction spread and recolonise.
Reptiles populate the land.
240 mya: Sea urchins (*Arkarua*) appear.
235 mya: Evolutionary split between dinosaurs and lizards.
Giant marine ichthyosaurs and plesiosaurs populate the seas.
First small dinosaurs such as *Coelophysis* appear on land.
Adelobasileus protomammal emerged (225 mya).
214 mya: Meteor impact, 100 km crater Manicouagan, Quebec, Canada (reference: Earth Impact Database).
205 mya: First evidence of mammals: *Morganucodon*.
201 mya: *Mass extinction caused by oceanic anoxic event killed 20% of all marine families.
Jurassic Period (199.6 to 145.5 mya)
Earth is warm. There is no polar ice.
Cycads, conifers and ginkgoes are the dominant plants.
Age of the dinosaurs.
Giant herbivores and vicious carnivores dominate the land.
Flying reptiles (Pterosaurs) appeared.
180 mya: North America separates from Africa.
167 mya: Meteor impact, 80 km crater Puchezh-Katunki, Russia (reference: Earth Impact Database).
166 mya: Evolutionary split of monotremes from primitive mammals.
150 mya: First birds such as *Archaeopteryx* appear.
148 mya: Evolutionary split between marsupial and eutherian mammals.

Table 13.1. (*cont.*)

145 mya: Meteor impact, 70 km crater Morokweng, South Africa (reference: Earth Impact Database).

Cretaceous Period (145.5 to 65.5 mya)

Period of Active Crust Plate Movements.

133 mya: Meteor impact, 55 km crater Tookoonooka, Australia (reference: Earth Impact Database).

125 mya: Africa and India separate from Antarctica.

Global warming event starts (120 mya).

Carbon dioxide levels were 550 to 590 ppm (Quan *et al.*, 2009).

Flowering plants (angiosperms) appeared.

110 mya: Crocodiles appeared.

South America breaks away from Africa (105 mya).

Formation of the Atlantic Ocean.

Earth has no polar ice.

Modern mammals and birds developed.

100 mya: Earth's magnetic field is three times stronger than today.

90 mya: Global warming event ends.

70 mya: Meteor impact, 65 km crater Kara, Russia (reference: Earth Impact Database).

68 mya: *Tyrannosaurus rex* thrived.

67 mya: Deccan Traps volcanic eruptions start in India and produce great volume of lava and gases.

65.5 mya: Meteor impact, 170 km crater Chicxulub, Yucatan, Mexico (reference: Earth Impact Database).

*Mass extinction of 80–90% of marine species and 85% of land species, including the dinosaurs.

Cenozoic Era (65.5 mya to today)

Palaeogene Period (65.5 to 23.03 mya)

Tertiary Period (65.5 to 2.58 mya)

Palaeocene Epoch (65.5 to 55.8 mya)

63 mya: End of Deccan Traps volcanic eruptions in India.

Appearance of placental mammals (marsupials, insectivores, lemuroids, creodonts).

Flowering plants become widespread.

60 mya: Earliest known ungulate (hoofed mammal).

Formation of the Rocky Mountains.

55.8 mya: Major global warming episode.

North Pole temperature averaged 23 °C (73.4 °F).

Carbon dioxide concentration was 2000 ppm.

Eocene Epoch (55.8 to 33.9 mya)

50 mya: India meets Asia forming the Himalayas.

45 mya: Australia separates from Antarctica.

Earth day is 24 hours long.

The Moon is 378 000 km from Earth (Williams, 2010).

Table 13.1. (*cont.*)

Modern mammals appear.
Rhinoceros, camels, early horses appear.
35.6 mya: Meteor impacts, 90 and 100 km craters Chesapeake Bay, Virginia, USA, and Popigai, Russia (reference: Earth Impact Database; Farley *et al.*, 1998).
34 mya: Global cooling creates permanent Antarctic ice sheet (Liu *et al.*, 2009).
Oligocene Epoch (33.9 to 23.03 mya)
Appearance of many grasses.
First elephants with trunks.
27.8 mya: La Garita, Colorado supervolcanic eruption.
Neogene Period (23.03 mya to today)
Miocene Epoch (23.03 to 5.3 mya)
African-Arabian plate joined to Asia.
14 mya: Antarctica separates from Australia and South America.
Circum-polar ocean circulation builds up Antarctic ice cap.
Warmer global climates.
First raccoons appear.
Drying of continental interiors.
Forests give way to grasslands.
6 mya: Upright walking (bipedal) hominids appear.
Pliocene Epoch (5.3 to 2.58 mya)
4.4 mya: Appearance of *Ardipithecus*, an early hominid genus.
4 mya: North and South America join at the Isthmus of Panama.
Animals and plants cross the new land bridge.
Ocean currents change in the newly isolated Atlantic Ocean.
3.9 mya: Appearance of *Australopithecus*, genus of hominids.
3.7 mya: *Australopithecus* hominids inhabit Eastern and Northern Africa.
3 mya: Formation of Arctic ice cap.
Accumulation of ice at the poles.
Climate became cooler and drier.
Spread of grasslands and savannahs.
Rise of long-legged grazing animals.
Quaternary Period (2.58 mya to today)
Pleistocene Epoch (2.58 mya to 11 400 yrs ago)
Several major episodes of global cooling, or glaciations.
2.4 mya: *Homo habilis* appeared.
2.1 mya: Yellowstone supervolcanic eruption.
2 mya: Tool-making humanoids emerge.
Beginning of the Stone Age.
1.7 mya: *Homo erectus* first moves out of Africa.
1.3 mya: Yellowstone supervolcanic eruption.
1.3 mya to 820 000 yrs ago: Sherwin Glaciation.
Presence of large land mammals and birds.

Table 13.1. (*cont.*)

700 000 yrs ago: Human and Neanderthal lineages start to diverge genetically.
680 000 to 620 000 yrs ago: Günz/Nebraskan glacial period.
640 000 yrs ago: Yellowstone supervolcanic eruption.
530 000 yrs ago: Development of speech in *Homo heidelbergensis* (Martínez *et al.*, 2008).
455 000 to 300 000 yrs ago: Mindel/Kansan glacial period.
400 000 yrs ago: Hominids hunt with wooden spears and use stone-cutting tools.
370 000 yrs ago: Human ancestors and Neanderthals are fully separate populations.
300 000 yrs ago: Hominids use controlled fires.
Neanderthal man spreads through Europe 230 000 yrs ago.
200 000 to 130 000 yrs ago: Riss/Illinoian glacial period.
160 000 yrs ago: *Homo sapiens* appeared.
Origin of human female lineage (Mitochondrial Eve).
125 000 yrs ago: Eemian stage or Riss/Würm interglacial period.
Hardwood forests grew above the Arctic Circle.
Melting ice sheets increased sea level by 6 metres (20 feet).
110 000 yrs ago: Start of Würm/Wisconsin glacial period.
105 000 yrs ago: Stone age humans forage for grass seeds such as sorghum.
80 000 yrs ago: Non-African humans interbreed with Neanderthals (Green *et al.*, 2010).
74 000 yrs ago: Toba volcanic eruption releases large volume of sulphur dioxide.
Homo sapiens reduced to about 10 000 individuals.
70 000 yrs ago: Tahoe glacial maximum.
Glaciers cover Canada and northern USA.
60 000 yrs ago: Oldest male ancestor of modern humans (Atlas of the Human Journey, 2011).
46 000 yrs ago: Australia becomes arid, bush fires destroy habitat, and megafauna die off.
40 000 yrs ago: Cro-Magnon man appeared in Europe.
28 000 yrs ago: Neanderthals disappear from fossil record (Finlayson *et al.*, 2006).
26 500 yrs ago: Taupo supervolcanic eruption in New Zealand.
22 000 yrs ago: Tioga glacial maximum sea level was 130 m lower than today.
19 000 yrs ago: Antarctic sea ice starts melting (Stott, Timmermann & Thunell, 2007).
15 000 yrs ago: Bering land bridge between Alaska and Siberia allows human migration to America.
12 900 yrs ago: Explosion of comet over Canada (Kennett *et al.*, 2009) causes extinction of American megafauna such as the mammoth and sabretooth cat (*Smilodon*), as well as the end of Clovis culture.

Table 13.1. (*cont.*)

Fired pottery invented (12 000 yrs ago).
11 400 yrs ago: End of Würm/Wisconsin glacial period.
Sea level rises by 91 m (300 ft).
Holocene Epoch (11 400 years ago to today)
Development of agriculture.
Domestication of animals.
9000 yrs ago: Metal smelting started.
5500 yrs ago: Invention of the wheel.
5300 yrs ago: The Bronze Age.
5000 yrs ago: Development of writing.
4500 yrs ago: Pyramids of Giza.
3300 yrs ago: The Iron Age.
2230 yrs ago: Archimedes advances mathematics.
250 yrs ago: Start of the Industrial Revolution.
50 yrs ago: Space travel.
Artificial satellite orbits the Earth (1957).
Humans walk on the surface of the Moon (1969).

Mya, million years ago.
*The five major mass extinction events occurred during the terminal Ordovician (443 mya), Late Devonian (374 mya), terminal Permian called the 'Great Dying' (251 mya), terminal Triassic (201 mya), and terminal Cretaceous called the K/T event (65.5 mya).
Data © 2006–2008 Antonio Zamora, taken from the *Scientific Psychic website* at this URL: www.scientificpsychic.com/index.html. Reproduced by permission of A. Zamora.

REFERENCES

These references include Internet URLs or DOI URLs. The acronym DOI stands for Digital Object Identifier, which uniquely identifies where an electronic document (or other electronic object) can be found on the Internet and remains fixed. Other information about a document may change over time, including where to find it, but its DOI name will not change and will always direct you to the original electronic document. To access one of these references enter the DOI URL into your browser and you will be taken to the document on the website of the original publisher. Almost always you will have free access to the abstract or summary of the article, but if your institution maintains a subscription to the products of that publisher you may be able to download the complete text of the article. Save the downloaded document to your hard disk to build your own reprint collection.

Alpermann, T., Rüdel, K., Rüger, R., *et al.* (2010). Polymersomes containing iron sulfide (FeS) as primordial cell model for the investigation of energy providing redox reactions. *Origins of Life and Evolution of Biospheres*, 41: 103–119. DOI: http://dx.doi.org/10.1007/s11084-010-9223-0.

Andrews-Hanna, J. C., Zuber, M. T. & Banerdt, W. B. (2008). The Borealis basin and the origin of the martian crustal dichotomy. *Nature*, 453: 1212–1215. DOI: http://dx.doi.org/10.1038/nature07011.

Arnaud-Haond, S., Duarte, C. M., Diaz-Almela, E., *et al.* (2012). Implications of extreme life span in clonal organisms: millenary clones in meadows of the threatened seagrass *Posidonia oceanica*. *PLoS ONE*, 7: e30454. DOI: http://dx.doi.org/10.1371/journal.pone.0030454.

Arrhenius, S. (1908). *Worlds in the Making: The Evolution of the Universe* (translated by H. Borns). New York: Harper & Row. URL: www.archive.org/details/worldsinmakingevooarrhrich.

Atkins, J. F., Gesteland, R. F. & Cech, T. R. (2010). *RNA Worlds: From Life's Origins to Diversity in Gene Regulation*. Cold Spring Harbor, NY: Cold Spring Harbor Laboratory Press. ISBN: 0879699469, 9780879699468.

Atlas of the Human Journey (2011). *The Genographic Project of the National Geographic Society*. URL: https://genographic.nationalgeographic.com/genographic/atlas.html.

Bahadur, K., Ranganayaki, S. & Santamaria, L. (1958). Photosynthesis of amino-acids from paraformaldehyde involving the fixation of nitrogen in the presence of colloidal molybdenum oxide as catalyst. *Nature*, **182**: 1668. DOI: http://dx.doi.org/10.1038/1821668a0.

Baltscheffsky, H., Blomberg, C., Liljenström, H., Lindahl, B. I. B. & Århem, P. (1997). On the origin and evolution of life: an introduction. *Journal of Theoretical Biology*, **187**: 453–459. DOI: http://dx.doi.org/10.1006/jtbi.1996.0380.

Baross, J. A. & Hoffman, S. E. (1985). Submarine hydrothermal vents and associated gradient environments as sites for the origin and evolution of life. *Origins of Life*, **15**: 327–345. DOI: http://dx.doi.org/10.1007/BF01808177.

Belbruno, E. & Gott, J. R. (2005). Where did the moon come from? *The Astronomical Journal*, **129**: 1724–1745. DOI: http://dx.doi.org/10.1086/427539.

Belloche, A., Garrod, R. T., Müller, H. S. P., *et al.* (2009). Increased complexity in interstellar chemistry: detection and chemical modeling of ethyl formate and n-propyl cyanide in Sagittarius B2(N). *Astronomy & Astrophysics*, **499**: 215–232. DOI: http://dx.doi.org/10.1051/0004-6361/200811550.

Benner, S. A. (2010). Defining life. *Astrobiology*, **10**: 1021–1030. DOI: http://dx.doi.org/10.1089/ast.2010.0524.

Bernstein, M. P. (2006). Prebiotic materials from on and off the early Earth. *Philosophical Transactions of the Royal Society of London, Series B*, **361**: 1689–1702. DOI: http://dx.doi.org/10.1098/rstb.2006.1913.

Bills, B. G. & Ray, R. D. (1999). Lunar orbital evolution: a synthesis of recent results. *Geophysical Research Letters*, **26**: 3045–3048. DOI: http://dx.doi.org/10.1029/1999GL008348.

Blackwell, M. (2000). Terrestrial life – fungal from the start? *Science*, **289**: 1884–1885. DOI: http://dx.doi.org/10.1126/science.289.5486.1884.

Blank, J. G., Miller, G. H., Ahrens, M. J. & Winans, R. E. (2001). Experimental shock chemistry of aqueous amino acid solutions and the cometary delivery of prebiotic compounds. *Origins of Life and Evolution of the Biosphere*, **31**: 15–51. DOI: http://dx.doi.org/10.1023/A:1006758803255.

Boal, D. & Ng, R. (2010). Shape analysis of filamentous Precambrian microfossils and modern cyanobacteria. *Paleobiology*, **36**: 555–572. DOI: http://dx.doi.org/10.1666/08096.1.

Bobe, R. & Behrensmeyer, A. K. (2004). The expansion of grassland ecosystems in Africa in relation to mammalian evolution and the origin of the genus *Homo*. *Palaeogeography, Palaeoclimatology, Palaeoecology*, **207**: 399–420. DOI: http://dx.doi.org/10.1016/j.palaeo.2003.09.033.

Boyce, C. K., Hotton, C. L., Fogel, M. L., *et al.* (2007). Devonian landscape heterogeneity recorded by a giant fungus. *Geology*, **35**: 399–402. DOI: http://dx.doi.org/10.1130/G23384A.1.

Bracker, C. E. (1968). The ultrastructure and development of sporangia in *Gilbertella persicaria. Mycologia*, **60**: 1016–1067. DOI: http://dx.doi.org/10.2307/3757290.

Breaker, R. R. (2011). Riboswitches and the RNA world. *Cold Spring Harbor Perspectives in Biology*, **3**: a003566. DOI: http://dx.doi.org/10.1101/cshperspect.a003566.

Brinkmann, H. & Philippe, H. (1999). Archaea sister-group of Bacteria? Indications from tree reconstruction artifacts in ancient phylogenies. *Molecular Biology and Evolution*, **16**: 817–825. URL: http://mbe.oxfordjournals.org/content/16/6/817.abstract.

Buss, L. W. (1983). Evolution, development, and the units of selection. *Proceedings of the National Academy of Sciences of the United States of America*, **80**: 1387–1391. Stable URL: www.jstor.org/stable/13577.

(1987). *The Evolution of Individuality*. Princeton, NJ: Princeton University Press. ISBN: 0691084696, 9780691084695.

Butterfield, N. J. (2000). *Bangiomorpha pubescens* n. gen., n. sp.: implications for the evolution of sex, multicellularity, and the Mesoproterozoic/Neoproterozoic radiation of eukaryotes. *Paleobiology*, **26**: 386–404. DOI: http://dx.doi.org/10.1666/0094-8373(2000)026<0386:BPNGNS>2.0.CO;2.

(2005). Probable Proterozoic fungi. *Paleobiology*, **31**: 165–182. DOI: http://dx.doi.org/10.1666/0094-8373(2005)031<0165:PPF>2.0.CO;2.

Cady, S. L. (2001). Ancient microbes, extreme environments, and the origin of life. *Advances in Applied Microbiology*, **50**: 3–35. DOI: http://dx.doi.org/10.1016/S0065-2164(01)50002-7.

Cairns-Smith, A. G. (1982). *Genetic Takeover and the Mineral Origins of Life*. Cambridge, UK: Cambridge University Press. ISBN: 0521346827, 9780521346825.

Callahan, M. P., Smith, K. E., Cleaves, H. J., *et al.* (2011). Carbonaceous meteorites contain a wide range of extraterrestrial nucleobases. *Proceedings of the National Academy of Sciences of the United States of America*, **108**: 13995–13998. DOI: http://dx.doi.org/10.1073/pnas.1106493108.

Cami, J., Bernard-Salas, J., Peeters, E. & Malek, S. E. (2010). Detection of C_{60} and C_{70} in a young planetary nebula. *Science*, **329**: 1180. DOI: http://dx.doi.org/10.1126/science.1192035.

Canfield, D. E., Rosing, M. T. & Bjerrum, C. (2006). Early anaerobic metabolisms. *Philosophical Transactions of the Royal Society of London, Series B*, **361**: 1819–1836. DOI: http://dx.doi.org/10.1098/rstb.2006.1906.

Carballeira, N. M., Reyes, M., Sostre, A., *et al.* (1997). Unusual fatty acid compositions of the hyperthermophilic archaeon *Pyrococcus furiosus* and the bacterium *Thermotoga maritima. Journal of Bacteriology*, **179**: 2766–2768. URL: http://jb.asm.org/content/179/8/2766.full.pdf+html.

Carny, O. & Gazit, E. (2005). A model for the role of short self-assembled peptides in the very early stages of the origin of life. *FASEB Journal (The Journal of the*

Federation of American Societies for Experimental Biology), **19**: 1051–1055. DOI: http://dx.doi.org/10.1096/fj.04-3256hyp.

Casadevall, A. (2005). Fungal virulence, vertebrate endothermy, and dinosaur extinction: is there a connection? *Fungal Genetics and Biology*, **42**: 98–106. DOI: http://dx.doi.org/10.1016/j.fgb.2004.11.008.

Cavalier-Smith, T. (2006). Cell evolution and Earth history: stasis and revolution. *Philosophical Transactions of the Royal Society of London, Series B*, **361**: 969–1006. DOI: http://dx.doi.org/10.1098/rstb.2006.1842.

(2010a). Deep phylogeny, ancestral groups and the four ages of life. *Philosophical Transactions of the Royal Society of London, Series B*, **365**: 111–132. DOI: http://dx.doi.org/10.1098/rstb.2009.0161.

(2010b). Kingdoms Protozoa and Chromista and the eozoan root of the eukaryotic tree. *Biology Letters*, **6**: 342–345. DOI: http://dx.doi.org/10.1098/rsbl.2009.0948.

Chyba, C. & Sagan, C. (1992). Endogenous production, exogenous delivery and impact-shock synthesis of organic molecules: an inventory for the origins of life. *Nature*, **355**: 125–132. DOI: http://dx.doi.org/10.1038/355125a0.

Chyba, C., Brookshaw, T. P. & Sagan, C. (1990). Cometary delivery of organic molecules to the early Earth. *Science*, **249**: 366–373. DOI: http://dx.doi.org/10.1126/science.11538074.

Cockell, C., Corfield, R., Edwards, N. & Harris, N. (2008). *An Introduction to the Earth-Life System*. Cambridge and Milton Keynes: Cambridge University Press in association with the Open University. ISBN 9780521729536.

Cockell, C. S. (2004). Impact-shocked rocks: insights into Archean and extraterrestrial microbial habitats (and sites for prebiotic chemistry?). *Advances in Space Research*, **33**: 1231–1235. DOI: http://dx.doi.org/10.1016/j.asr.2003.06.027.

Cody, G. D., Boctor, N. Z., Filley, T. R., *et al.* (2000). Primordial carbonylated iron-sulfur compounds and the synthesis of pyruvate. *Science*, **289**: 1337–1340. DOI: http://dx.doi.org/10.1126/science.289.5483.1337.

Conway Morris, S. (2003). *Life's Solution: Inevitable Humans in a Lonely Universe*. Cambridge, UK: Cambridge University Press. ISBN: 0521603250, 9780521603256.

Crick, F. H. & Orgel, L. E. (1973). Directed panspermia. *Icarus*, **19**: 341–348. DOI: http://dx.doi.org/10.1016/0019-1035(73)90110-3.

Darwin, E. & Litchfield, H. (1915). *Emma Darwin V2: A Century of Family Letters, 1792–1896*. London: John Murray. Republished, 2010, by Kessinger Publishing, ISBN: 1166051900, 9781166051907 [and see: http://darwin-online.org.uk/contents.html].

Darwin, F. (ed.) (1887). *The Life and Letters of Charles Darwin, Including an Autobiographical Chapter, Edited by his Son, Francis Darwin in Three Volumes*. Volume 3 (see p. 18). London: John Murray. [and see: http://darwin-online.org.uk/contents.html].

Davies, P. (2006). *The Origin of Life*. Harmondsworth, UK: Penguin. ISBN: 978-0141013022.

Dawkins, R. (1986). *The Blind Watchmaker: Why the Evidence of Evolution Reveals a Universe Without Design*. London: W. W. Norton & Co. ISBN-10: 0393022161, ISBN-13: 978-0393022162.

(2006). *The God Delusion*. London: Transworld Publishers. ISBN-10: 055277331X, ISBN-13: 9780552773317.

Deamer, D. W. (1985). Boundary structures are formed by organic components of the Murchison carbonaceous chondrite. *Nature*, **317**: 792–794. DOI: http://dx.doi.org/10.1038/317792a0.

Deamer, D. W. & Weber, A. L. (2010). Bioenergetics and life's origins. *Cold Spring Harbor Perspectives in Biology*, **2**: a004929. DOI: http://dx.doi.org/10.1101/cshperspect.a004929.

Deamer, D. W., Dworkin, J. P., Sandford, S. A., Bernstein, M. P. & Allamandola, L. J. (2002). The first cell membranes. *Astrobiology*, **2**: 371–381. DOI: http://dx.doi.org/10.1089/153110702762470482.

Derenne, S., Robert, F., Skrzypczak-Bonduelle, A., *et al.* (2008). Molecular evidence for life in the 3.5 billion year old Warrawoona Chert. *Earth and Planetary Science Letters*, **272**: 476–480. DOI: http://dx.doi.org/10.1016/j.epsl.2008.05.014.

Dobson, C. M., Ellison, G. B., Tuck, A. F. & Vaida, V. (2000). Atmospheric aerosols. *Proceedings of the National Academy of Sciences of the United States of America*, **97**: 11864–11868. DOI: http://dx.doi.org/10.1073/pnas.200366897.

Dobson, D. P. & Brodholt, J. P. (2005). Subducted banded iron formations as a source of ultralow-velocity zones at the core-mantle boundary. *Nature*, **434**: 371–374. DOI: http://dx.doi.org/10.1038/nature03430.

Donaldson, D. J., Tervahattu, H., Tuck, A. F. & Vaida, V. (2004). Organic aerosols and the origin of life: an hypothesis. *Origins of Life and Evolution of the Biosphere*, **34**: 57–67. DOI: http://dx.doi.org/10.1023/B:ORIG.0000009828.40846.b3.

Donaldson, D. J., Tuck, A. F. & Vaida, V. (2002). The asymmetry of organic aerosol fission and prebiotic chemistry. *Origins of Life and Evolution of Biospheres*, **32**: 237–245. DOI: http://dx.doi.org/10.1023/A:1016575224538.

Dörfelt, H. & Schmidt, A. R. (2005). A fossil *Aspergillus* from Baltic amber. *Mycological Research*, **109**: 956–960. DOI: http://dx.doi.org/10.1017/S0953756205003497.

Dyson, F. J. (1999). *Origins of Life*. Cambridge, UK: Cambridge University Press. ISBN: 0521626684, 9780521626682.

Earth Impact Database (2011). University of New Brunswick, Fredericton, New Brunswick, Canada: Planetary and Space Science Centre. URL: www.passc.net/EarthImpactDatabase/index.html.

Ehrenfreund, P. & Cami, J. (2010). Cosmic carbon chemistry: from the interstellar medium to the early Earth. *Cold Spring Harbor Perspectives in Biology*, **2**:(12): a002097. DOI: http://dx.doi.org/10.1101/cshperspect.a002097.

Ehrenfreund, P. & Foing, B. H. (2010). Fullerenes and cosmic carbon. *Science*, **329**: 1159–1160. DOI: http://dx.doi.org/10.1126/science.1194855.

Ehrenfreund, P., Charnley, S. B. & Botta, O. (2005). A voyage from dark clouds to the early Earth. In *Astrophysics of Life* (Proceedings of the Space Telescope Science Institute Symposium held in Baltimore, Maryland, May 6–9, 2002), ed. M. Livio, I. N. Reid & W. B. Sparks, pp. 1–22. Cambridge, UK: Cambridge University Press.

Ehrenfreund, P., Irvine, W., Becker, L., *et al.* & ISSI-Team 'Prebiotic matter in space' (2002). Astrophysical and astrochemical insights into the origin of life. *Reports on Progress in Physics*, **65**: 1427–1487. DOI: http://dx.doi.org/10.1088/0034–4885/65/10/202.

Eichler, J. & Adams, M. W. W. (2005). Posttranslational protein modification in Archaea. *Microbiology and Molecular Biology Reviews*, **69**: 393–425. DOI: http://dx.doi.org/10.1128/MMBR.69.3.393–425.2005.

Eigen, M. & Schuster, P. (1977). The hypercycle: a principle of natural self-organization. Part A: Emergence of the hypercycle. *Naturwissenschaften*, **64**: 541–565. DOI: http://dx.doi.org/10.1007/BF00450633.

(1978a). The hypercycle: a principle of natural self-organization. Part B: The abstract hypercycle. *Naturwissenschaften*, **65**: 7–41. DOI: http://dx.doi.org/10.1007/BF00420631.

(1978b). The hypercycle: a principle of natural self-organization. Part C: The realistic hypercycle. *Naturwissenschaften*, **65**: 341–369. DOI: http://dx.doi.org/10.1007/BF00439699.

Farley, K. A., Montanari, A., Shoemaker, E. M. & Shoemaker, C. S. (1998). Geochemical evidence for a comet shower in the late Eocene. *Science*, **280**: 1250–1253. DOI: http://dx.doi.org/10.1126/science.280.5367.1250.

Fenchel, T. (2002). *The Origin & Early Evolution of Life*. Oxford, UK: Oxford University Press. ISBN: 0198525338, 9780198525332.

Ferguson, B. A., Dreisbach, T. A., Parks, C. G., Filip, G. M. & Schmitt, C. L. (2003). Coarse-scale population structure of pathogenic *Armillaria* species in a mixed-conifer forest in the Blue Mountains of northeast Oregon. *Canadian Journal of Forest Research*, **33**: 612–623. DOI: http://dx.doi.org/10.1139/x03–065.

Finlayson, C., Pacheco, F. G., Rodríguez-Vidal, J., *et al.* (2006). Late survival of Neanderthals at the southernmost extreme of Europe. *Nature*, **443**: 850–853. DOI: http://dx.doi.org/10.1038/nature05195.

Flemming, H.-C. & Wingender, J. (2010). The biofilm matrix. *Nature Reviews Microbiology*, **8**: 623–633. DOI: http://dx.doi.org/10.1038/nrmicro2415.

Flemming, H.-C., Wingender, J. & Szewzyk, U. (2011). *Biofilm Perspectives*. Volume 5 of Springer Series on Biofilms. Berlin and Heidelberg: Springer-Verlag GmbH & Co. KG. ISBN: 3642199399, 9783642199394.

Follmann, H. & Brownson, C. (2009). Darwin's warm little pond revisited: from molecules to the origin of life. *Naturwissenschaften*, **96**: 1265–1292. DOI: http://dx.doi.org/10.1007/s00114-009-0602-1.

Fox, S. W. (1980). Life from an orderly cosmos. *Naturwissenschaften*, **67**: 576–581. DOI: http://dx.doi.org/10.1007/BF00396536.

Fraser, C. L. & Folsome, C. E. (1975). Exponential kinetics of formation or organic microstructures. *Origins of Life and Evolution of Biospheres*, **6**: 429–433. DOI: http://dx.doi.org/10.1007/BF01130345.

Gánti, T. (2003). *The Principles of Life* (with a commentary by James Griesemer & Eörs Szathmáry). Oxford, UK: Oxford University Press. ISBN: 0198507267, 9780198507260.

García-Hernández, D. A., Manchado, A., García-Lario, P., *et al.* (2010). Formation of fullerenes in H-containing planetary nebulae. *Astrophysical Journal Letters*, **724**: L39–L43. DOI: http://dx.doi.org/10.1088/2041–8205/724/1/L39.

Gilbert, W. (1986). The RNA world. *Nature*, **319**: 618. DOI: http://dx.doi.org/10.1038/319618a0.

Glass, N. L., Rasmussen, C., Roca, M. G. & Read, N. D. (2004). Hyphal homing, fusion and mycelial interconnectedness. *Trends in Microbiology*, **12**: 135–141. DOI: http://dx.doi.org/10.1016/j.tim.2004.01.007.

Glavin, D. P. & Dworkin, J. P. (2009). Enrichment in L-isovaline by aqueous alteration on CI and CM meteorite parent bodies. *Proceedings of the National Academy of Sciences of the United States of America*, **106**: 5487–5492. DOI: http://dx.doi.org/10.1073/pnas.0811618106.

Gogarten-Boekels, M., Hilario, E. & Gogarten, P. (1995). The effects of heavy meteorite bombardment on the early evolution – the emergence of the three domains. *Origins of Life and Evolution of the Biosphere*, **25**: 251–264. DOI: http://dx.doi.org/10.1007/BF01581588.

Gold, T. (1992). The deep, hot biosphere. *Proceedings of the National Academy of Sciences of the United States of America*, **89**: 6045–6049. URL: www.pnas.org/content/89/13/6045.abstract.

Goldreich, P. (1966). History of the Lunar orbit. *Reviews of Geophysics*, **4**: 411–439. DOI: http://dx.doi.org/10.1029/RG004i004p00411.

Green, R. E., Krause, J., Briggs, A. W., *et al.* (2010). A draft sequence of the Neandertal genome. *Science*, **328**: 710–722. DOI: http://dx.doi.org/10.1126/science.1188021.

Haldane, J. B. S. (1929). The origin of life. *Rationalist Annual*, **3**: 3–10.

Harding, M. W., Marques, L. L. R., Howard, R. J. & Olson, M. E. (2009). Can filamentous fungi form biofilms? *Trends in Microbiology*, **17**: 475–480. DOI: http://dx.doi.org/10.1016/j.tim.2009.08.007.

Hartman, H. (1975). Speculations on the origin and evolution of metabolism. *Journal of Molecular Evolution*, **4**: 359–370. DOI: http://dx.doi.org/10.1007/BF01732537.

Hartmann, W. K. & Davis, D. R. (1975). Satellite-sized planetesimals and lunar origin. *Icarus*, **24**: 504–515. DOI: http://dx.doi.org/10.1016/0019-1035(75)90070-6.

Hazen, R. M. (2005). *Genesis: The Scientific Quest for Life's Origin*. Washington, DC: Joseph Henry Press. ISBN: 0309094321, 9780309094320.

Hereward, F. V. & Moore, D. (1979). Polymorphic variation in the structure of aerial sclerotia of *Coprinus cinereus*. *Journal of General Microbiology*, **113**: 13–18. DOI: http://dx.doi.org/10.1099/00221287-113-1-13.

Hibbett, D. S., Grimaldi, D. & Donoghue, M. J. (1995). Cretaceous mushrooms in amber. *Nature*, **377**: 487. DOI: http://dx.doi.org/10.1038/377487a0.

Hobbie, E. A. & Boyce, C. K. (2010). Carbon sources for the Palaeozoic giant fungus *Prototaxites* inferred from modern analogues. *Proceedings of the Royal Society, Series B*, **277**: 2149–2156. DOI: http://dx.doi.org/10.1098/rspb.2010.0201.

Horneck, G. (1996). Exobiology. In *Biological and Medical Research in Space: An Overview of Life Sciences Research in Microgravity*, ed. D. Moore, P. Bie & H. Oser, Chapter 7. Berlin, Heidelberg & New York: Springer-Verlag. ISBN: 354060636X, 9783540606369.

(1999). European activities in exobiology in Earth orbit: results and perspectives. *Advances in Space Research*, **23**: 381–386. DOI: http://dx.doi.org/10.1016/S0273-1177(99)00061-7.

Hoyle, F. & Wickramasinghe, N. C. (1982). *Proofs that Life is Cosmic.* Colombo, Sri Lanka: Colombo Government Press. URL: www.panspermia.org/proofslifeiscosmic.pdf.

Huber, C. & Wächtershäuser, G. (1997). Activated acetic acid by carbon fixation on (Fe,Ni)S under primordial conditions. *Science*, **276**: 245–247. DOI: http://dx.doi.org/10.1126/science.276.5310.245.

Hueber, F. M. (2001). Rotted wood-alga-fungus: the history and life of *Prototaxites* Dawson 1859. *Review of Paleobotany and Palynology*, **116**: 123–148. DOI: http://dx.doi.org/10.1016/S0034-6667(01)00058-6.

Hull, D. L. (1988). *Science as a Process: An Evolutionary Account of the Social and Conceptual Development of Science.* Chicago, IL: University of Chicago Press. ISBN: 0226360512, 9780226360515.

Jacobs, B. F. (2004). Palaeobotanical studies from tropical Africa: relevance to the evolution of forest, woodland and savannah biomes. *Philosophical Transactions of the Royal Society of London, Series B*, **359**: 1573–1583. DOI: http://dx.doi.org/10.1098/rstb.2004.1533.

Javaux, E. J., Knoll, A. H. & Walter, M. R. (2001). Morphological and ecological complexity in early eukaryotic ecosystems. *Nature*, **412**: 66–69. DOI: http://dx.doi.org/10.1038/35083562.

Johnson, A. P., Cleaves, H. J., Dworkin, J. P., *et al.* (2008). The Miller volcanic spark discharge experiment. *Science*, **322**: 404. DOI: http://dx.doi.org/10.1126/science.1161527.

Jutzi, M. & Asphaug, E. (2011). Forming the lunar farside highlands by accretion of a companion moon. *Nature*, **476**: 69–72. DOI: http://dx.doi.org/10.1038/nature10289.

Karatan, E. & Watnick, P. (2009). Signals, regulatory networks, and materials that build and break bacterial biofilms. *Microbiology and Molecular Biology Reviews*, **73**: 310–347. DOI: http://dx.doi.org/10.1128/MMBR.00041-08.

Kasting, J. F. (1993). Earth's early atmosphere. *Science*, **259**: 920–926. DOI: http://dx.doi.org/10.1126/science.11536547.

Kasting, J. F. & Howard, M. T. (2006). Atmospheric composition and climate on the early Earth. *Philosophical Transactions of the Royal Society of London, Series B*, **361**: 1733–1742. DOI: http://dx.doi.org/10.1098/rstb.2006.1902.

Keeling, P. J., Burger, G., Durnford, D. G., *et al.* (2005). The tree of eukaryotes. *Trends in Ecology and Evolution*, **20**: 670–676. DOI: http://dx.doi.org/10.1016/j.tree.2005.09.005.

Kelley, D. S., Carson, J. A., Blackman, D. K., *et al.* & the At3-60 Shipboard Party (2001). An off-axis hydrothermal vent field discovered near the Mid-Atlantic Ridge at 30°N. *Nature*, **412**: 145–149. DOI: http://dx.doi.org/10.1038/35084000.

Kennett, D. J., Kennett, J. P., West, A., *et al.* (2009). Nanodiamonds in the Younger Dryas boundary sediment layer. *Science*, **323**: 94. DOI: http://dx.doi.org/10.1126/science.1162819.

Koshland, D. E. (2002). The seven pillars of life. *Science*, **295**: 2215–2216. DOI: http://dx.doi.org/10.1126/science.1068489.

Kwok, S. (2004). The synthesis of organic and inorganic compounds in evolved stars. *Nature*, **430**: 985–991. DOI: http://dx.doi.org/10.1038/nature02862.

Lane, N. (2010). *Life Ascending: The Ten Great Inventions of Evolution*. London: Profile Books Ltd. ISBN: 9781861978189.

Lane, N., Allen, J. F. & Martin, W. (2010). How did LUCA make a living? Chemiosmosis in the origin of life. *BioEssays*, **32**: 271–280. DOI: http://dx.doi.org/10.1002/bies.200900131.

Lasaga, A. C., Holland, H. D. & Dwyer, M. J. (1971). Primordial oil slick. *Science*, **174**: 53–55. DOI: http://dx.doi.org/10.1126/science.174.4004.53.

Lazcano, A. (2010). Which way to life? *Origins of Life and Evolution of Biospheres*, **40**: 161–167. DOI: http://dx.doi.org/10.1007/s11084-010-9195-0.

Lazcano, A. & Miller, S. L. (1996). The origin and early evolution of life: prebiotic chemistry, the pre-RNA world, and time. *Cell*, **85**: 793–798. DOI: http://dx.doi.org/10.1016/S0092-8674(00)81263-5.

Lessie, P. E. & Lovett, J. S. (1968). Ultrastructural changes during sporangium formation and zoospore differentiation in *Blastocladiella emersonii*. *American Journal of Botany*, **55**: 220–236. URL: www.jstor.org/stable/2440456.

Liu, Z., Pagani, M., Zinniker, D., *et al.* (2009). Global cooling during the Eocene-Oligocene climate transition. *Science*, **323**: 1187–1190. DOI: http://dx.doi.org/10.1126/science.1166368.

Lopez, P., Forterre, P. & Philippe, H. (1999). The root of the tree of life in the light of the covarion model. *Journal of Molecular Evolution*, **49**: 496–508. DOI: http://dx.doi.org/10.1007/PL00006572.

Luisi, P. L. (1998). About various definitions of life. *Origins of Life and Evolution of the Biosphere*, **28**: 613–622. DOI: http://dx.doi.org/10.1023/A:1006517315105.

(2006). *The Emergence of Life: From Chemical Origins to Synthetic Biology*. Cambridge, UK: Cambridge University Press. ISBN: 0521821177, 9780521821179.

Lunine, J. I. (2006). Physical conditions on the early Earth. *Philosophical Transactions of the Royal Society of London, Series B*, **361**: 1721–1731. DOI: http://dx.doi.org/10.1098/rstb.2006.1900.

Lurquin, P. F. (2003). *The Origins of Life and the Universe*. New York: Columbia University Press. ISBN: 0231126557, 9780231126557.

Mackie, R. I. (2002). Mutualistic fermentative digestion in the gastrointestinal tract: diversity and evolution. *Integrative & Comparative Biology*, **42**: 319–326. DOI: http://dx.doi.org/10.1093/icb/42.2.319.

Maher, K. A. & Stevenson, D. J. (1988). Impact frustration of the origin of life. *Nature*, **331**: 612–614. DOI: http://dx.doi.org/10.1038/331612a0.

Margulis, L. (2004). Serial endosymbiotic theory (SET) and composite individuality: transition from bacterial to eukaryotic genomes. *Microbiology Today*, **31**: 172–174. DOI: www.socgenmicrobiol.org.uk/pubs/micro_today/pdf/110406.pdf.

Martin, W. & Russell, M. J. (2007). On the origin of biochemistry at an alkaline hydrothermal vent. *Philosophical Transactions of the Royal Society of London, Series B*, **362**: 1887–1926. DOI: http://dx.doi.org/10.1098/rstb.2006.1881.

Martin, W., Rotte, C., Hoffmeister, M., *et al.* (2003). Early cell evolution, eukaryotes, anoxia, sulfide, oxygen, fungi first (?), and a tree of genomes revisited. *International Union of Biochemistry and Molecular Biology: Life*, **55**: 193–204. DOI: http://dx.doi.org/10.1080/1521654031000141231.

Martínez, I., Arsuaga, J. L., Quam, R., *et al.* (2008). Human hyoid bones from the middle Pleistocene site of the Sima de los Huesos (Sierra de Atapuerca, Spain). *Journal of Human Evolution*, **54**: 118–124. DOI: http://dx.doi.org/10.1016/j.jhevol.2007.07.006.

Martins, Z. (2011). Organic chemistry of carbonaceous meteorites. *Elements*, **7**: 35–40. DOI: http://dx.doi.org/10.2113/gselements.7.1.35.

Maynard Smith, J. & Szathmáry, E. (1999). *The Origins of Life: From the Birth of Life to the Origin of Language*. Oxford, UK: Oxford University Press. ISBN: 019286209X.

Melosh, H. J. (1988). The rocky road to panspermia. *Nature*, **332**: 687–688. DOI: http://dx.doi.org/10.1038/332687a0.

Miller, S. L. (1953). Production of amino acids under possible primitive Earth conditions. *Science*, **117**: 528–529. DOI: http://dx.doi.org/10.1126/science.117.3046.528.

Miller, S. L. & Bada, J. L. (1988). Submarine hot springs and the origin of life. *Nature*, **334**: 609–611. DOI: http://dx.doi.org/10.1038/334609a0.

Moore, D. (1981). Developmental genetics of *Coprinus cinereus*: genetic evidence that carpophores and sclerotia share a common pathway of initiation. *Current Genetics*, **3**: 145–150. DOI: http://dx.doi.org/10.1007/BF00365718.

(1998). *Fungal Morphogenesis*. New York: Cambridge University Press. ISBN: 0521552958, 9780521552950. DOI: http://dx.doi.org/10.1017/CBO9780511529887.

(2000). *Slayers, Saviors, Servants and Sex: An Exposé of Kingdom Fungi*. New York: Springer-Verlag. ISBN-10: 0387951016, ISBN-13: 9780387951010. URL: www.springer.com/life+sciences/microbiology/book/978-0-387-95098-3.

(2005). Principles of mushroom developmental biology. *International Journal of Medicinal Mushrooms*, **7**: 79–102. DOI: http://dx.doi.org/10.1615/IntJMedMushr.v7.i12.90.

Moore, D. & Meškauskas, A. (2006). A comprehensive comparative analysis of the occurrence of developmental sequences in fungal, plant and animal genomes. *Mycological Research*, **110**: 251–256. DOI: http://dx.doi.org/10.1016/j.mycres.2006.01.003.

Moore, D. & Pöder, R. (2006). Are your children taught anything about fungi at school? *Sydowia*, **58**: 1–2. URL: www.sydowia.at/syd58-1/T1-Moore.htm.

Moore, D., Pöder, R., Molitoris, H. P., *et al.* (2006). Crisis in teaching future generations about fungi. *Mycological Research*, **110**: 626–627. DOI: http://dx.doi.org/10.1016/j.mycres.2006.05.005.

Moore, D., Robson, G. D. & Trinci, A. P. J. (2011). *21st Century Guidebook to Fungi*. Cambridge, UK: Cambridge University Press. ISBN: 9780521186957.

Newsom, H. E. & Taylor, S. R. (1989). Geochemical implications of the formation of the Moon by a single giant impact. *Nature*, **338**: 29–34. DOI: http://dx.doi.org/10.1038/338029a0.

Nimmo, F., Hart, S. D., Korycansky, D. G. & Agnor, C. B. (2008). Implications of an impact origin for the martian hemispheric dichotomy. *Nature*, **453**: 1220–1223. DOI: http://dx.doi.org/10.1038/nature07025.

Nisbet, E. (2000). Palaeobiology: the realms of Archaen life. *Nature*, **405**: 625–626. DOI: http://dx.doi.org/10.1038/35015187.

Nisbet, E., Zahnle, K., Gerasimov, M. V., *et al.* (2007). Creating habitable zones, at all scales, from planets to mud micro-habitats, on Earth and on Mars. *Space Science Reviews*, **24**: 79–121. DOI: http://dx.doi.org/10.1007/978-0-387-74288-5_4.

Oparin, A. I. (1957a). *The Origin of Life on the Earth*. 3rd edn, translated by Ann Synge. London: Oliver and Boyd. ASIN: B001I4YTQI.

(1957b). Biochemical processes in the simplest structures. In *International Symposium on the Origin of Life on the Earth*, ed. A. I. Oparin, A. Braunstein, N. Gelman, G. Deborin & A. Passynsky, pp. 221–228. Moscow: The Publishing House of the Academy of Sciences of the USSR.

Orgel, L. E. (1998). The origin of life: a review of facts and speculations. *Trends in Biochemical Sciences (TIBS)*, **23**: 491–495. DOI: http://dx.doi.org/10.1016/S0968-0004(98)01300-0.

(2004). Prebiotic adenine revisited: eutectics and photochemistry. *Origins of Life and Evolution of Biospheres*, **34**: 361–369. DOI: http://dx.doi.org/10.1023/B:ORIG.0000029882.52156.c2.

Oró, J. (1961). Comets and the formation of biochemical compounds on the primitive Earth. *Nature*, **190**: 389–390. DOI: http://dx.doi.org/10.1038/190389a0.

Parker, E. T., Cleaves, H. J., Dworkin, J. P., *et al.* (2011). Primordial synthesis of amines and amino acids in a 1958 Miller H_2S-rich spark discharge experiment. *Proceedings of the National Academy of Sciences of the United States of America*, **108**: 5526–5531. DOI: http://dx.doi.org/10.1073/pnas.1019191108.

Pennisi, E. (2004). The birth of the nucleus. *Science*, **305**: 766–768. DOI: http://dx.doi.org/10.1126/science.305.5685.766.

Penny, D. & Poole, A. (1999). The nature of the last universal common ancestor. *Current Opinion in Genetics & Development*, **9**: 672–677. DOI: http://dx.doi.org/10.1016/S0959-437X(99)00020-9.

Pirozynski, K. A. (1976). Fungal spores in fossil record. *Biological Memoirs*, **1**: 104–120.

Quan, C., Sun, C., Sun, Y. & Sun, G. (2009). High resolution estimates of paleo-CO_2 levels through the Campanian (Late Cretaceous) based on *Ginkgo* cuticles. *Cretaceous Research*, **30**: 424–428. DOI: http://dx.doi.org/10.1016/j.cretres.2008.08.004.

Read, N. D., Fleißner, A, Roca, M. G. & Glass, N. L. (2010). Hyphal fusion. In *Cellular and Molecular Biology of Filamentous Fungi*, ed. K. A. Borkovich & D. J. Ebbole, pp. 260–273. Washington, DC: American Society for Microbiology Press. ISBN-10: 1555814735, ISBN-13: 978-1555814731.

Read, N. D., Lichius, A., Shoji, J.-Y. & Goryachev, A. B. (2009). Self-signalling and self-fusion in filamentous fungi. *Current Opinion in Microbiology*, **12**: 608–615. DOI: http://dx.doi.org/10.1016/j.mib.2009.09.008.

Redecker, D., Kodner, R. & Graham, L. E. (2000). Glomalean fungi from the Ordovician. *Science*, **289**: 1920–1921. DOI: http://dx.doi.org/10.1126/science.289.5486.1920.

Reynolds, T. B. & Fink, G. R. (2001). Bakers' yeast, a model for fungal biofilm formation. *Science*, **291**: 878–881. DOI: http://dx.doi.org/10.1126/science.291.5505.878.

Rikkinen, J., Dörfelt, H., Schmidt, A. R. & Wunderlich, J. (2003). Sooty moulds from European Tertiary amber, with notes on the systematic position of *Rosaria* ('Cyanobacteria'). *Mycological Research*, **107**: 251–256. DOI: http://dx.doi.org/10.1017/S0953756203007330.

Riquelme, M. & Bartnicki-García, S. (2008). Advances in understanding hyphal morphogenesis: ontogeny, phylogeny and cellular localization of chitin synthases. *Fungal Biology Reviews*, **22**: 56–70. DOI: http://dx.doi.org/10.1016/j.fbr.2008.05.003.

Riquelme, M., Bartnicki-García, S., González-Prieto, J. M., *et al.* (2007). Spitzenkörper localization and intracellular traffic of green fluorescent protein-labeled CHS-3 and CHS-6 chitin synthases in living hyphae of *Neurospora crassa*. *Eukaryotic Cell*, **6**: 1853–1864. DOI: http://dx.doi.org/10.1128/EC.00088–07.

Robb, F. T. & Clark, D. S. (1999). Adaptation of proteins from hyperthermophiles to high pressure and high temperature. *Journal of Molecular Biotechnology*, **1**: 101–105. URL: www.horizonpress.com/jmmb/v1/v1n1/15.pdf.

Robertson, M. P. & Joyce, G. F. (2011). The origins of the RNA world. *Cold Spring Harbor Perspectives in Biology*, **3**: a003608v2. DOI: http://dx.doi.org/10.1101/cshperspect.a003608.

Ruiz-Bermejo, M., Menor-Salván, C., Osuna-Esteban, S. & Veintemillas-Verdaguer, S. (2007). The effects of ferrous and other ions on the abiotic formation of biomolecules using aqueous aerosols and spark discharges. *Origins of Life and Evolution of Biospheres*, **37**: 507–521. DOI: http://dx.doi.org/10.1007/s11084–007–9107–0.

Russell, M. J. (2010). The hazy details of early Earth's atmosphere. *Science*, **330**: 754. DOI: http://dx.doi.org/10.1126/science.330.6005.754-a.

Russell, M. J. & Hall, A. J. (2002). From geochemistry to biochemistry: chemiosmotic coupling and transition element clusters in the onset of life and photosynthesis. *The Geochemical News, Newsletter of the Geochemical Society*, **113**: 6–12. URL: www.geochemsoc.org/downloads/gn113.pdf.

Russell, M. J., Daniel, R. M., Hall, A. J. & Sherringham, J. (1994). A hydrothermally precipitated catalytic iron sulphide membrane as a first step toward life. *Journal of Molecular Evolution*, **39**: 231–243. DOI: http://dx.doi.org/10.1007/BF00160147.

Sagan, C. & Chyba, C. (1997). The early faint sun paradox: organic shielding of ultraviolet-labile greenhouse gases. *Science*, **276**: 1217–1221. DOI: http://dx.doi.org/10.1126/science.276.5316.1217.

Sanders, W. B. (2001). Lichens: interface between mycology and plant morphology. *BioScience*, **51**: 1025–1035. DOI: http://dx.doi.org/10.1641/0006–3568(2001)051[1025:LTIBMA]2.0.CO;2.

Schopf, J. W. (1993). Microfossils of the early Archean Apex Chert: new evidence of the antiquity of life. *Science*, **260**: 640–646. DOI: http://dx.doi.org/10.1126/science.260.5108.640.

Schrödinger, E. (1944). *What is Life?* [reprinted 1992, with Mind and Matter and Autobiographical Sketches]. Cambridge, UK: Cambridge University Press. ISBN: 0521427088, 9780521427081.

Schultz, T.R. & Brady, S.G. (2008). Major evolutionary transitions in ant agriculture. *Proceedings of the National Academy of Sciences of the United States of America*, **105**: 5435–5440. DOI: http://dx.doi.org/10.1073/pnas.0711024105.

Scott, E.C. (2009). *Evolution vs. Creationism: An Introduction.* 2nd revised edn. Berkeley and Los Angeles, CA: University of California Press. ISBN-10: 0520261879, ISBN-13: 978–0520261877.

Simoncini, E., Russell, M.J. & Kleidon, A. (2011). Modeling free energy availability from Hadean hydrothermal systems to the first metabolism. *Origins of Life and Evolution of Biospheres*. Epub ahead of print. DOI: http://dx.doi.org/10.1007/s11084–011–9251–4.

Singh, R., Shivaprakash, M.R. & Chakrabarti, A. (2011). Biofilm formation by zygomycetes: quantification, structure and matrix composition. *Microbiology*. Online: ahead of print. DOI: http://dx.doi.org/10.1099/mic.0.048504–0.

Sorokin, Y.I. (1957). The evolution of chemosynthesis. In *International Symposium on the Origin of Life on the Earth*, ed. A.I. Oparin, A. Braunstein, N. Gelman, G. Deborin & A. Passynsky, pp. 368–375. Moscow: The Publishing House of the Academy of Sciences of the USSR.

Steinberg, G. (2007). Hyphal growth: a tale of motors, lipids, and the Spitzenkörper. *Eukaryotic Cell*, **6**: 351–360. DOI: http://dx.doi.org/10.1128/EC.00381–06.

Steinberg, G. & Schuster, M. (2011). The dynamic fungal cell. *Fungal Biology Reviews*, **25**: 14–37. DOI: http://dx.doi.org/10.1016/j.fbr.2011.01.008.

Stetter, K.O. (2006). Hyperthermophiles in the history of life. *Philosophical Transactions of the Royal Society of London, Series B*, **361**: 1837–1843. DOI: http://dx.doi.org/10.1098/rstb.2006.1907.

Stott, L., Timmermann, A. & Thunell, R. (2007). Southern hemisphere and deep-sea warming led deglacial atmospheric CO_2 rise and tropical warming. *Science*, **318**: 435–438. DOI: http://dx.doi.org/10.1126/science.1143791.

Strömberg, C.A.E. & Feranec, R.S. (2004). The evolution of grass-dominated ecosystems during the late Cenozoic. *Palaeogeography, Palaeoclimatology, Palaeoecology*, **207**: 199–201. DOI: http://dx.doi.org/10.1016/j.palaeo.2004.01.017.

Sudarsan, N., Barrick, J.E. & Breaker, R.R. (2010). Metabolite-binding RNA domains are present in the genes of eukaryotes. *RNA*, **9**: 644–647. DOI: http://dx.doi.org/10.1261/rna.5090103.

Sutherland, I.W. (2001). The biofilm matrix: an immobilized but dynamic microbial environment. *Trends in Microbiology*, **9**: 222–227. DOI: http://dx.doi.org/10.1016/S0966–842X(01)02012–1.

Taylor, J.W., Jacobson, D.J. & Fisher, M.C. (1999). The evolution of asexual fungi: reproduction, speciation and classification. *Annual Review of Phytopathology*, **37**: 197–246. DOI: http://dx.doi.org/10.1146/annurev.phyto.37.1.197.

Taylor, T.N., Hass, H. & Kerp, H. (1997). A cyanolichen from the Lower Devonian Rhynie Chert. *American Journal of Botany*, **84**: 992–1004. Stable URL: www.jstor.org/stable/2446290.

Taylor, T.N., Klavins, S.D., Krings, M., *et al.* (2004). Fungi from the Rhynie chert: a view from the dark side. *Transactions of the Royal Society of Edinburgh: Earth Sciences*, **94**: 457–473. DOI: http://dx.doi.org/10.1017/S026359330000081X.

Taylor, T. N., Krings, M. & Kerp, H. (2006). *Hassiella monospora* gen. et sp. nov., a microfungus from the 400 million year old Rhynie chert. *Mycological Research*, **110**: 628–632. DOI: http://dx.doi.org/10.1016/j.mycres.2006.02.009.

Taylor, T. N., Taylor, E. L., Decombeix, A. -L., *et al.* (2010). The enigmatic Devonian fossil *Prototaxites* is not a rolled-up liverwort mat: comment on the paper by Graham *et al.* (AJB 97: 268–275). *American Journal of Botany*, **97**: 1074–1078. DOI: http://dx.doi.org/10.3732/ajb.1000047.

Thaddeus, P. (2006). The prebiotic molecules observed in the interstellar gas. *Philosophical Transactions of the Royal Society of London, Series B*, **361**: 1681–1687. DOI: http://dx.doi.org/10.1098/rstb.2006.1897.

Tucker, B. J. & Breaker, R. R. (2005). Riboswitches as versatile gene control elements. *Current Opinion in Structural Biology*, **15**: 342–348. DOI: http://dx.doi.org/10.1016/j.sbi.2005.05.003.

Urey, H. C. (1952). On the early chemical history of the earth and the origin of life. *Proceedings of the National Academy of Sciences of the United States of America*, **38**: 351–363. URL: www.pnas.org/content/38/4/351.short.

Vajda, V. & McLoughlin, S. (2004). Fungal proliferation at the Cretaceous-Tertiary boundary. *Science*, **303**: 1489. DOI: http://dx.doi.org/10.1126/science.1093807.

van Wyhe, J. (2010). 'Almighty God! What a wonderful discovery!': Did Charles Darwin really believe life came from space? *Endeavour*, **34**: 95–103. DOI: http://dx.doi.org/10.1016/j.endeavour.2010.07.003.

Vijh, U. P., Witt, A. N. & Gordon, K. D. (2005). Small polycyclic aromatic hydrocarbons in the Red Rectangle. *Astrophysical Journal*, **619**: 368–378. DOI: http://dx.doi.org/10.1086/426498.

Visscher, H., Brinkuis, H., Dilcher, D. L., *et al.* (1996). The terminal Paleozoic fungal event: evidence of terrestrial ecosystem destabilization and collapse. *Proceedings of the National Academy of Sciences of the United States of America*, **93**: 2155–2158. URL: www.jstor.org/stable/38482.

von Frese, R. R. B., Potts, L. V., Wells, S. B., *et al.* (2009). GRACE gravity evidence for an impact basin in Wilkes Land, Antarctica. *Geochemistry, Geophysics, and Geosystems*, **10**: Q02014. DOI: http://dx.doi.org/10.1029/2008GC002149.

Wächtershäuser, G. (1988). Before enzymes and templates: theory of surface metabolism. *Microbiological Reviews*, **52**: 452–484. URL: www.ncbi.nlm.nih.gov/pmc/articles/PMC373159/.

(1992). Groundworks for an evolutionary biochemistry: the iron-sulphur world. *Progress in Biophysics and Molecular Biology*, **58**: 85–201. DOI: http://dx.doi.org/10.1016/0079-6107(92)90022-X.

(2000). Life as we don't know it. *Science*, **289**: 1307–1308. DOI: http://dx.doi.org/10.1126/science.289.5483.1307.

(2006). From volcanic origins of chemoautotrophic life to Bacteria, Archaea and Eukarya. *Philosophical Transactions of the Royal Society of London, Series B*, **361**: 1787–1808. DOI: http://dx.doi.org/10.1098/rstb.2006.1904.

Ward, P. D. (2006). Impact from the deep. *Scientific American*, **295**: 64–71. URL: www.scientificamerican.com/article.cfm?id=impact-from-the-deep.

Waters, H., Butler, R. D. & Moore, D. (1975). Structure of aerial and submerged sclerotia of *Coprinus lagopus*. *New Phytologist*, **74**: 199–205. DOI: http://dx.doi.org/10.1111/j.1469-8137.1975.tb02606.x.

Waters, H., Moore, D. & Butler, R. D. (1975). Morphogenesis of aerial sclerotia of *Coprinus lagopus*. *New Phytologist*, **74**: 207–213. DOI: http://dx.doi.org/10.1111/j.1469-8137.1975.tb02607.x.

Wellman, C. H. & Gray, J. (2000). The microfossil record of early land plants. *Philosophical Transactions of the Royal Society of London, Series B*, **355**: 717–732. URL: www.jstor.org/stable/3066802.

Westall, F., deRonde, C. E. J., Southam, G., *et al.* (2006). Implications of a 3.472–3.333 Gyr-old subaerial microbial mat from the Barberton greenstone belt, South Africa for the UV environmental conditions on the early Earth. *Philosophical Transactions of the Royal Society of London, Series B*, **361**: 1857–1875. DOI: http://dx.doi.org/10.1098/rstb.2006.1896.

Whittaker, R. H. (1969). New concepts of kingdoms of organisms. *Science*, **163**: 150–160. DOI: http://dx.doi.org/10.1126/science.163.3863.150.

Wickramasinghe, N. C. (2010). The astrobiological case for our cosmic ancestry. *International Journal of Astrobiology*, **9**: 119–129. DOI: http://dx.doi.org/10.1017/S1473550409990413.

Williams, D. R. (2010). Moon Fact Sheet, NASA Goddard Space Flight Center. URL: http://nssdc.gsfc.nasa.gov/planetary/factsheet/moonfact.html.

Winkler, W. C., Nahvi, A., Roth, A., Collins, J. A. & Breaker, R. R. (2004). Control of gene expression by a natural metabolite-responsive ribozyme. *Nature*, **428**: 281–286. DOI: http://dx.doi.org/10.1038/nature02362.

Woese, C. R. (1987). Bacterial evolution. *Microbiological Reviews*, **51**: 221–271. URL: www.ncbi.nlm.nih.gov/pmc/articles/PMC373105/.

Woese, C. R., Kandler, O. & Wheels, M. L. (1990). Towards a natural system of organisms: proposal for the domains Archaea, Bacteria and Eucarya. *Proceedings of the National Academy of Sciences of the United States of America*, 87: 4576–4579. URL: www.jstor.org/stable/2354364.

Wood, W. B. (ed.) (1988). *The Nematode, Caenorhabditis elegans*. Cold Spring Harbor, NY: Cold Spring Harbor Laboratory Press. ISBN: 9780879694333.

Yuan, X., Xiao, S. & Taylor, T. N. (2005). Lichen-like symbiosis 600 million years ago. *Science*, **308**: 1017–1020. DOI: http://dx.doi.org/10.1126/science.1111347.

Zahnle, K., Arndt, N., Cockell, C., *et al.* (2007). Emergence of a habitable planet. *Space Science Reviews*, **129**: 35–78. DOI: http://dx.doi.org/10.1007/s11214-007-9225-z.

Zahnle, K., Schaefer, L. & Fegley, B. (2010). Earth's earliest atmospheres. *Cold Spring Harbor Perspectives in Biology*, **2**: a004895. DOI: http://dx.doi.org/10.1101/cshperspect.a004895.

Zaug, A. J. & Cech, T. R. (1986). The intervening sequence RNA of *Tetrahymena* is an enzyme. *Science*, **231**: 470–475. DOI: http://dx.doi.org/10.1126/science.3941911.

INDEX